Springer Series in

SOLID-STATE SCIENCES 159

Springer Series in
SOLID-STATE SCIENCES

The Springer Series in Solid-State Sciences consists of fundamental scientific books prepared by leading researchers in the field. They strive to communicate, in a systematic and comprehensive way, the basic principles as well as new developments in theoretical and experimental solid-state physics.

Please view available titles in *Springer Series in Solid-State Sciences* on series homepage http://www.springer.com/series/682

Alexandre S. Alexandrov
Jozef T. Devreese

Advances in Polaron Physics

With 43 Figures

Springer

Professor Dr. Alexandre S. Alexandrov
Loughborough University
Dept. Physics
Loughborough
United Kingdom LE11 3TU
E-mail: a.s.alexandrov@lboro.ac.uk

Professor Dr. Jozef T. Devreese
Universiteit Antwerpen
Groenenborgerlaan 171
2020 Antwerpen
Belgium
E-mail: jozef.devreese@ua.ac.be

Springer Series in Solid-State Sciences ISSN 0171-1873
ISBN 978-3-642-01895-4 e-ISBN 978-3-642-01896-1
DOI 10.1007/978-3-642-01896-1
Springer Heidelberg Dordrecht London New York

Library of Congress Control Number: 2009929173

Cover design: SPi Publisher Services

Printed on acid-free paper

Springer is part of Springer Science+Business Media (www.springer.com)

Preface

While basic features of polarons were well recognized a long time ago and have been described in a number of review papers and textbooks, interest in the role of electron–phonon interactions and polaron dynamics in different materials has recently gone through a vigorous revival. Electron–phonon interactions have been shown to be relevant in many inorganic and organic semiconductors and polymers, colossal magnetoresistance oxides, and transport through nanowires and quantum dots also often depends on vibronic displacements of ions. These interactions presumably play a role in high-temperature superconductors as well. The continued interest in polarons extends beyond the physical description of advanced materials. The field has been a testing ground for analytical, semi-analytical, and numerical techniques, such as path integrals, strong-coupling perturbation expansion, advanced variational methods, exact diagonalization, Quantum Monte Carlo, and other techniques. This book reviews some recent developments in the field of polarons, starting with the basics and covering a number of active directions of research. Single- and multipolaron theories have offered more insight into colossal magnetoresistance and in a broad spectrum of physical properties of structures with reduced dimension and dimensionality such as transport, optical absorption, Raman scattering, photoluminescence, magneto-optics, etc. While nobody – at present – has a final theory of high-temperature superconductivity, we discuss one alternative (polaronic) route.

We have benefited from discussions with many experts in the field. While writing this book and at some earlier stage of our studies, discussions with A. Andreev, S. Aubry, I. Bozovic, A. Bratkovsky, F. Brosens, P. Edwards, J. Bonca, V. Cataudella, G. De Filippis, Yu. Firsov, V. Fomin, R. Evrard, H. Fehske, V. Gladilin, J. Hague, M. Hohenadler, G. Iadonisi, V. Kabanov, E. Kartheuser, S. Klimin, P. Kornilovitch, L. Lemmens, W. von der Linden, D. Mihailovic, A. Mishchenko, G. Papadopoulos, F. Peeters, J. Samson, M. Stoneham, J. Tempere, and P. Zhao were especially helpful.

Loughborough
Antwerp
June 2009

Alexandre S. Alexandrov
Jozef T. Devreese

Contents

1

Introduction

1.1 First-Principle Hamiltonian

Charge carriers in inorganic and organic matter interact with ion vibrations. The corresponding electron–phonon interaction (EPI) dominates transport and other properties of many poor metals and semiconductors. EPI causes also phase transformations, including superconductivity. When EPI is sufficiently strong, electron Bloch states are affected even in the normal phase. Phonons are also affected by conduction electrons. In doped insulators, including the advanced materials discussed in this review, *bare* phonons are well defined in insulating parent compounds, but microscopic separation of electrons and phonons is not so straightforward in metals and heavily doped insulators [1], where the Born and Oppenheimer [2] and density functional [3,4] methods are used. Here, we have to start with the first-principle Hamiltonian describing conduction electrons and ions coupled by the Coulomb forces:

$$H = -\sum_i \frac{\nabla_i^2}{2m_\mathrm{e}} + \frac{e^2}{2}\sum_{i\neq i'} \frac{1}{|\mathbf{r}_i - \mathbf{r}_{i'}|} - Ze^2\sum_{ij}\frac{1}{|\mathbf{r}_i - \mathbf{R}_j|} + \frac{Z^2e^2}{2}\sum_{j\neq j'}\frac{1}{|\mathbf{R}_j - \mathbf{R}_{j'}|} - \sum_j \frac{\nabla_j^2}{2M}, \tag{1.1}$$

where $\mathbf{r}_i, \mathbf{R}_j$ are the electron and ion coordinates, respectively, $i = 1, \ldots, N_\mathrm{e}$; $j = 1, \ldots, N$; $\nabla_i = \partial/\partial\mathbf{r}_i$, $\nabla_j = \partial/\partial\mathbf{R}_j$, Ze is the ion charge, and M is the ion mass. The system is neutral, $N_\mathrm{e} = ZN$. The inner electrons are strongly coupled to the nuclei and follow their motion, so the ions can be considered as rigid charges. To account for their "high-energy" electron degrees of freedom, we can replace the elementary charge in (1.1) by $e/\sqrt{\epsilon}$, where ϵ is the high-frequency dielectric constant, or introduce an electron–ion "pseudopotential" [5] instead of the bare Coulomb electron–nuclear interaction.

A.S. Alexandrov and J.T. Devreese, *Advances in Polaron Physics*, Springer Series in Solid-State Sciences 159,
DOI: 10.1007/978-3-642-01896-1_1,

1.2 Local Density Approximation

One cannot solve the corresponding Schrödinger equation perturbatively because the Coulomb interaction is strong. The ratio of the characteristic Coulomb energy to the kinetic energy is $r_s = m_e e^2/(4\pi n_e/3)^{1/3} \approx 1$ for the electron density $n_e = ZN = 10^{23}\,\mathrm{cm}^{-3}$ (here and further, we take the volume of the system as $V = 1$, unless specified otherwise, and $\hbar = c = k_B = 1$). However, one can take advantage of the small value of the adiabatic ratio $m_e/M < 10^{-3}$. Ions are heavy and the amplitudes of their vibrations, $\langle|\mathbf{u}|\rangle \simeq \sqrt{1/M\omega_D}$, near equilibrium positions $\mathbf{R}_0 \equiv \mathbf{l}$ are much smaller than the lattice constant $a = N^{-1/3}$:

$$\frac{\langle|\mathbf{u}|\rangle}{a} \approx \left(\frac{m_e}{Mr_s}\right)^{1/4} \ll 1. \tag{1.2}$$

In this estimate, we take the characteristic vibration frequency ω_D of the order of the ion plasma frequency $\omega_0 = \sqrt{4\pi NZ^2e^2/M}$. Hence, one can expand the Hamiltonian in powers of $|\mathbf{u}|$.

Any further progress requires a simplifying physical idea, which commonly is to approach the ground state of the many-electron system via a one-electron picture. In the framework of the local density approximation (LDA), the Coulomb electron–electron interaction is replaced by an effective one-body potential $\mathcal{V}(\mathbf{r})$:

$$\mathcal{V}(\mathbf{r}) = -Ze^2\sum_j \frac{1}{|\mathbf{r}-\mathbf{R}_j|} + e^2\int \mathrm{d}\mathbf{r}'\frac{n(\mathbf{r}')}{|\mathbf{r}-\mathbf{r}'|} + \mu_{ex}[n(\mathbf{r})], \tag{1.3}$$

where $\mu_{ex}[n(\mathbf{r})]$ is the exchange interaction, usually calculated numerically or expressed as $\mu_{ex}[n(\mathbf{r})] = -\beta n(\mathbf{r})^{1/3}$ with a constant β in the simplest approximation. Here, $\mathcal{V}(\mathbf{r})$ is a functional of the electron density $n(\mathbf{r}) = \sum_s\langle\Psi_s^\dagger(\mathbf{r})\Psi_s(\mathbf{r})\rangle$, where $\Psi_s(\mathbf{r})$ annihilates the electron with spin s and coordinate $\mathbf{r}$. As a result, in second quantization for electrons the Hamiltonian takes the form

$$H = H_e + H_{ph} + H_{e\text{-}ph} + H_{e\text{-}e}, \tag{1.4}$$

where

$$H_e = \sum_s \int \mathrm{d}\mathbf{r}\Psi_s^\dagger(\mathbf{r})\left[-\frac{\nabla^2}{2m_e} + V(\mathbf{r})\right]\Psi_s(\mathbf{r}) \tag{1.5}$$

is the electron energy in a periodic crystal field $V(\mathbf{r}) = \sum_\mathbf{l} v(\mathbf{r}-\mathbf{l})$, which is $\mathcal{V}(\mathbf{r})$ calculated at $\mathbf{R}_j = \mathbf{l}$ with the periodic electron density $n^{(0)}(\mathbf{r}+\mathbf{l}) = n^{(0)}(\mathbf{r})$:

$$H_{ph} = \sum_\mathbf{l}\left[-\frac{\nabla_\mathbf{u}^2}{2M} + \mathbf{u}_\mathbf{l}\cdot\frac{\partial}{\partial\mathbf{l}}\int \mathrm{d}\mathbf{r}n^{(0)}(\mathbf{r})V(\mathbf{r})\right] + \frac{1}{2}\sum_{\mathbf{l},\mathbf{m},\alpha,\beta} u_{\mathbf{l}\alpha}u_{\mathbf{m}\beta}D_{\alpha\beta}(\mathbf{l}-\mathbf{m}) \tag{1.6}$$

is the vibration energy. Here, $\alpha, \beta = x, y, z$ and

$$D_{\alpha\beta}(\mathbf{l}-\mathbf{m}) = \frac{\partial^2}{\partial l_\alpha \partial m_\beta}\left[\frac{Z^2e^2}{2}\sum_{\mathbf{l}'\neq\mathbf{m}'}\frac{1}{|\mathbf{l}'-\mathbf{m}'|} + \int d\mathbf{r} n^{(0)}(\mathbf{r})V(\mathbf{r})\right] \tag{1.7}$$

is the dynamic matrix. EPI is given by

$$\begin{aligned} H_{\text{e-ph}} &= \sum_{\mathbf{l}} \mathbf{u}_{\mathbf{l}} \cdot \frac{\partial}{\partial \mathbf{l}} \int d\mathbf{r} \left[\sum_s \Psi_s^\dagger(\mathbf{r})\Psi_s(\mathbf{r}) - n^{(0)}(\mathbf{r})\right] V(\mathbf{r}) \\ &+ \frac{1}{2}\sum_{\mathbf{l},\mathbf{m},\alpha,\beta} u_{\mathbf{l}\alpha}u_{\mathbf{m}\beta}\frac{\partial^2}{\partial l_\alpha \partial m_\beta}\int d\mathbf{r}\left[\sum_s \Psi_s^\dagger(\mathbf{r})\Psi_s(\mathbf{r}) - n^{(0)}(\mathbf{r})\right] V(\mathbf{r}), \end{aligned} \tag{1.8}$$

and the electron–electron correlations are described by

$$\begin{aligned} H_{\text{e-e}} &= \frac{1}{2}\int d\mathbf{r}\int d\mathbf{r}' \frac{e^2}{|\mathbf{r}-\mathbf{r}'|}\left[\sum_{ss'}\Psi_s^\dagger(\mathbf{r})\Psi_{s'}^\dagger(\mathbf{r}')\Psi'_{s'}(\mathbf{r}')\Psi_s(\mathbf{r})\right] \\ &- \int d\mathbf{r}\left[\int d\mathbf{r}'\frac{e^2 n^{(0)}(\mathbf{r}')}{|\mathbf{r}-\mathbf{r}'|} + \mu_{\text{ex}}[n^{(0)}(\mathbf{r})]\right]\sum_s \Psi_s^\dagger(\mathbf{r})\Psi_s(\mathbf{r}) \\ &+ \frac{Z^2e^2}{2}\sum_{\mathbf{l}\neq\mathbf{m}}\frac{1}{|\mathbf{l}-\mathbf{m}|}. \end{aligned} \tag{1.9}$$

We include the electrostatic repulsive energy of the nuclei in $H_{\text{e-e}}$, so that the average of $H_{\text{e-e}}$ is zero in the Hartree approximation.

1.3 Electron–Phonon Interaction in the Bloch Representation

The vibration Hamiltonian H_{ph} is a quadratic form and therefore can be diagonalized using the linear canonical transformation for the displacement operators:

$$\begin{aligned} \mathbf{u}_{\mathbf{l}} &= \sum_{\mathbf{q},\nu}\frac{\mathbf{e}_{\mathbf{q}\nu}}{\sqrt{2NM\omega_{\mathbf{q}\nu}}}d_{\mathbf{q}\nu}\exp(i\mathbf{q}\cdot\mathbf{l}) + \text{H.c.}, \\ \frac{\partial}{\partial \mathbf{u}_{\mathbf{l}}} &= \sum_{\mathbf{q},\nu}\mathbf{e}_{\mathbf{q}\nu}\sqrt{\frac{M\omega_{\mathbf{q}\nu}}{2N}}d_{\mathbf{q}\nu}\exp(i\mathbf{q}\cdot\mathbf{l}) - \text{H.c.}, \end{aligned} \tag{1.10}$$

where $\mathbf{q}$ is the phonon momentum, $d_{\mathbf{q}\nu}$ is the phonon (Bose) annihilation operator, $\mathbf{e}_{\mathbf{q}\nu}$ and $\omega_{\mathbf{q}\nu}$ are the unit polarization vector and the phonon frequency, respectively, of the phonon mode ν, and H.c. is the Hermitian conjugate. Then, H_{ph} takes the following form

$$H_{\mathrm{ph}} = \sum_{\mathbf{q},\nu} \omega_{\mathbf{q}\nu}(d^{\dagger}_{\mathbf{q}\nu} d_{\mathbf{q}\nu} + 1/2), \tag{1.11}$$

if the eigenfrequencies $\omega_{\mathbf{q}\nu}$ and the eigenstates $\mathbf{e}_{\mathbf{q}\nu}$ satisfy

$$M\omega^2_{\mathbf{q}\nu} e^{\alpha}_{\mathbf{q}\nu} = \sum_{\beta} D^{\alpha\beta}_{\mathbf{q}} e^{\beta}_{\mathbf{q}\nu}, \tag{1.12}$$

and

$$\sum_{\mathbf{q}} e^{*\alpha}_{\mathbf{q}\nu} e^{\beta}_{\mathbf{q}\nu} = N\delta_{\alpha\beta}. \tag{1.13}$$

The last equation and the bosonic commutation rules $[d_{\mathbf{q}\nu} d^{\dagger}_{\mathbf{q}'\nu'}] = \delta_{\nu\nu'}\delta_{\mathbf{q}\mathbf{q}'}$ follow from $(\partial/\partial u^{\alpha}_{\mathbf{l}})u^{\beta}_{\mathbf{l}} - u^{\beta}_{\mathbf{l}}(\partial/\partial u^{\alpha}_{\mathbf{l}}) = \delta_{\alpha\beta}$. Here

$$D^{\alpha\beta}_{\mathbf{q}} = \sum_{\mathbf{m}} \exp(\mathrm{i}\mathbf{q}\cdot\mathbf{m}) D_{\alpha\beta}(\mathbf{m}) \tag{1.14}$$

is the Fourier transform of the second derivative of the ion potential energy. The first derivative in (1.6) is zero in crystals with a center of symmetry. Different solutions of (1.12) are classified with the phonon branch (mode) quantum number ν, which is $1, 2, 3$ for a simple lattice and $1, \ldots, 3k$ for a complex lattice with k ions per unit cell.

The periodic part of the Hamiltonian H_{e} is diagonal in the Bloch representation:

$$\Psi_s(\mathbf{r}) = \sum_{n,\mathbf{k}} \psi_{n\mathbf{k}s}(\mathbf{r}) c_{n\mathbf{k}s}, \tag{1.15}$$

where $c_{n\mathbf{k}s}$ are the fermion annihilation operators. The Bloch function obeys the Schrödinger equation

$$\left(-\frac{\nabla^2}{2m_{\mathrm{e}}} + V(\mathbf{r})\right) \psi_{n\mathbf{k}s}(\mathbf{r}) = E_{n\mathbf{k}s}\psi_{n\mathbf{k}s}(\mathbf{r}). \tag{1.16}$$

One-particle states are sorted with the momentum $\mathbf{k}$ in the Brillouin zone, band index n, and spin s. The solution of this equation allows us to calculate the periodic electron density $n^{(0)}(\mathbf{r})$, which determines the crystal field potential $V(\mathbf{r})$. LDA can explain the shape of the Fermi surface of wide-band metals and gaps in narrow-gap semiconductors. A spin-polarized version of LDA can explain a variety of properties of many magnetic materials. However, this is not the case for narrow d- and f-band metals and oxides (and other ionic lattices), where the electron–phonon interaction and Coulomb correlations are strong. These materials display much less band dispersion and wider gaps compared with the first-principle band structure calculations.

Using the phonon and electron annihilation and creation operators, the Hamiltonian is finally written as

$$H = H_{\mathrm{e}} + H_{\text{e–ph}} + H_{\text{e–e}}, \tag{1.17}$$

where

$$H_{\mathrm{e}} = \sum_{\mathbf{k},n,s} \xi_{n\mathbf{k}s} c^{\dagger}_{n\mathbf{k}s} c_{n\mathbf{k}s} + \sum_{\mathbf{q},\nu} \omega_{\mathbf{q}\nu}(d^{\dagger}_{\mathbf{q}\nu} d_{\mathbf{q}\nu} + 1/2) \tag{1.18}$$

describes independent Bloch electrons and phonons, $\xi_{n\mathbf{k}s} = E_{n\mathbf{k}s} - \mu$ is the band energy spectrum with respect to the chemical potential μ. The part of EPI, which is linear in the phonon operators, can be written as

$$H_{\text{e–ph}} = \frac{1}{\sqrt{2N}} \sum_{\mathbf{k},\mathbf{q},n,n',\nu,s} \gamma_{nn'}(\mathbf{q},\mathbf{k},\nu)\omega_{\mathbf{q}\nu} c^{\dagger}_{n\mathbf{k}s} c_{n\mathbf{k}-\mathbf{q}s} d_{\mathbf{q}\nu} + \text{H.c.}, \tag{1.19}$$

where

$$\gamma_{nn'}(\mathbf{q},\mathbf{k},\nu) = -\frac{N}{M^{1/2}\omega^{3/2}_{\mathbf{q}\nu}} \int \mathrm{d}\mathbf{r}\,(\mathbf{e}_{\mathbf{q}\nu} \cdot \nabla v(\mathbf{r}))\,\psi^{*}_{n\mathbf{k}s}(\mathbf{r})\psi_{n'\mathbf{k}-\mathbf{q}s}(\mathbf{r}) \tag{1.20}$$

is the dimensionless matrix element. If we restrict the summations over $\mathbf{q}$ and $\mathbf{k}$ to the first Brillouin zone of the crystal, then $H_{\text{e–ph}}$ should also include the summation over reciprocal lattice vectors $\mathbf{G}$ of umklapp scattering contributions where $\mathbf{q}$ is replaced by $\mathbf{q} + \mathbf{G}$. The terms of $H_{\text{e–ph}}$ which are quadratic and of higher orders in the phonon operators are small. They play a role only for those phonons which are not coupled with electrons by the linear interaction (1.19).

The electron–electron correlation energy of a homogeneous electron system is often written as

$$H_{\text{e–e}} = \frac{1}{2}\sum_{\mathbf{q}} V_{\mathrm{c}}(\mathbf{q})\rho^{\dagger}_{\mathbf{q}}\rho_{\mathbf{q}}, \tag{1.21}$$

where $V_{\mathrm{c}}(\mathbf{q})$ is a matrix element, which is zero for $\mathbf{q} = 0$ because of electroneutrality and

$$\rho^{\dagger}_{\mathbf{q}} = \sum_{\mathbf{k},s} c^{\dagger}_{\mathbf{k}s} c_{\mathbf{k}+\mathbf{q}s} \tag{1.22}$$

is the density fluctuation operator. H should also include a random potential in doped semiconductors and amorphous metals, which might affect the EPI matrix element [6].

1.4 Electron–Phonon Interaction in the Wannier (Site) Representation

For the purpose of this review, we mostly confine our discussions to a single-band approximation with the EPI matrix element $\gamma_{nn}(\mathbf{q},\mathbf{k},\nu) = \gamma(\mathbf{q})$ depending only on the momentum transfer $\mathbf{q}$. The approximation allows for qualitative and in many cases quantitative descriptions of essential polaronic effects in advanced materials. Nevertheless, there might be degenerate atomic orbitals in solids coupled to local molecular-type Jahn–Teller distortions, where one has to consider multiband electron energy structures (see Sect. 3.6).

Quantitative calculations of the matrix element in the whole region of momenta have to be performed from pseudopotentials [1, 5]. On the other hand, one can parameterize EPI rather than to compute it from first principles in many physically important cases [7]. There are three most important interactions in doped semiconductors, which are polar coupling to optical phonons (i.e., the Fröhlich EPI), deformation potential coupling to acoustical phonons, and the local (Holstein) EPI with molecular-type vibrations in complex lattices. While the matrix element is ill defined in metals, the bare phonons $\omega_{\mathbf{q}\nu}$ and the electron band structure $E_{n\mathbf{k}}$ are well defined in doped semiconductors, which have their parent dielectric compounds. Here, the effect of carriers on the crystal field and on the dynamic matrix is small while the carrier density is much less than the atomic one. Hence, one can use the band structure and the crystal field of parent insulators to calculate the matrix element in doped semiconductors. The interaction constant $\gamma(\mathbf{q})$ has different q-dependence for different phonon branches. In the long-wavelength limit ($q \ll \pi/a$), $\gamma(\mathbf{q}) \propto q^n$, where $n = -1, 0$ and $n = -1/2$ for polar optical, molecular ($\omega_{\mathbf{q}} = \omega_0$) and acoustic ($\omega_{\mathbf{q}} \propto q$) phonons, respectively. Not only q dependence is known, but also the absolute values of $\gamma(\mathbf{q})$ are well parameterized in this limit. For example in polar semiconductors, the interaction of two slow electrons at some distance r is found as (3.40)

$$v(r) = V_{\mathrm{c}}(r) - \frac{1}{N}\sum_{\mathbf{q}} |\gamma(\mathbf{q})|^2 \omega_{\mathbf{q}} \mathrm{e}^{\mathrm{i}\mathbf{q}\cdot\mathbf{r}}. \tag{1.23}$$

The Coulomb repulsion in a rigid lattice is $V_{\mathrm{c}}(r) = e^2/\epsilon r$ and $|\gamma(\mathbf{q})|^2 = \gamma^2/q^2$, so that

$$v(r) = \frac{e^2}{\epsilon r} - \frac{\gamma^2 \omega_0}{4\pi r}. \tag{1.24}$$

On the other hand, the Coulomb repulsion can be expressed as $v(r) = e^2/\epsilon_0 r$, where the static dielectric constant, ϵ_0, describes screening by both core electrons and ions. The high frequency, ϵ, and static, ϵ_0, dielectric constants are measurable: ϵ is the square of the refractive index and ϵ_0 is the low-frequency dielectric function measured by putting the solid in a capacitor. Hence, the matrix element of the Fröhlich interaction depends only on the dielectric constants and the optical phonon frequency ω_0 as

$$|\gamma(\mathbf{q})|^2 = \frac{4\pi e^2}{\kappa \omega_0 q^2}, \tag{1.25}$$

where $\kappa = (\epsilon^{-1} - \epsilon_0^{-1})^{-1}$. If the crystal lacks an inversion center to be piezoelectric, there is EPI with piezoelectric acoustic phonons with an anysotropic matrix element, which also contribute to a coulombic-like attraction of two polarons [8].

To get a better insight into physical constraints of the above approximation, let us transform the Bloch states to the real space or Wannier states using the canonical linear transformation of the electron operators (here we follow [9]):

$$c_i = \frac{1}{\sqrt{N}} \sum_{\mathbf{k}} \mathrm{e}^{\mathrm{i}\mathbf{k}\cdot\mathbf{m}} c_{\mathbf{k}s}, \tag{1.26}$$

where $i = (\mathbf{m}, s)$ includes both site $\mathbf{m}$ and spin s quantum numbers. In this site (Wannier) representation, the electron kinetic energy takes the following form

$$H_{\mathrm{e}} = \sum_{i,j} t(\mathbf{m} - \mathbf{n}) \delta_{ss'} c_i^{\dagger} c_j, \tag{1.27}$$

where

$$t(\mathbf{m}) = \frac{1}{N} \sum_{\mathbf{k}} E_{\mathbf{k}} \mathrm{e}^{\mathrm{i}\mathbf{k}\cdot\mathbf{m}}$$

is the "bare" hopping integral. Here, $j = (\mathbf{n}, s')$ and $E_{\mathbf{k}}$ is the Bloch band dispersion in the rigid lattice.

The electron–phonon interaction and the Coulomb correlations acquire simple forms in the Wannier representation, if their matrix elements in the momentum representation depend only on the momentum transfer $\mathbf{q}$:

$$H_{\mathrm{e\text{-}ph}} = \sum_{\mathbf{q},i} \omega_{\mathbf{q}} \hat{n}_i \left[u_i(\mathbf{q}) d_{\mathbf{q}} + \mathrm{H.c.} \right], \tag{1.28}$$

$$H_{\mathrm{e\text{-}e}} = \frac{1}{2} \sum_{i \neq j} V_{\mathrm{c}}(\mathbf{m} - \mathbf{n}) \hat{n}_i \hat{n}_j. \tag{1.29}$$

Here

$$u_i(\mathbf{q}) = \frac{1}{\sqrt{2N}} \gamma(\mathbf{q}) \mathrm{e}^{\mathrm{i}\mathbf{q}\cdot\mathbf{m}} \tag{1.30}$$

and

$$V_{\mathrm{c}}(\mathbf{m}) = \frac{1}{N} \sum_{\mathbf{q}} V_{\mathrm{c}}(\mathbf{q}) \mathrm{e}^{\mathrm{i}\mathbf{q}\cdot\mathbf{m}} \tag{1.31}$$

are the matrix elements of the electron–phonon and Coulomb interactions, respectively, in the Wannier representation for electrons, and $\hat{n}_i = c_i^{\dagger} c_i$ is the density operator.

We see that taking the interaction matrix element depending only on the momentum transfer, one neglects the terms in the electron–phonon and

Coulomb interactions, which are proportional to the overlap integrals of the Wannier orbitals on different sites. This approximation is justified for narrow-band materials, where the electron bandwidth is less than the characteristic magnitude of the crystal field potential. In the Wannier representation, the Hamiltonian becomes

$$\begin{aligned} H = & \sum_{i,j} t(\mathbf{m}-\mathbf{n})\delta_{ss'} c_i^\dagger c_j + \sum_{\mathbf{q},i} \omega_{\mathbf{q}} \hat{n}_i \left[u_i(\mathbf{q}) d_{\mathbf{q}} + \text{H.c.}\right] \\ & + \frac{1}{2}\sum_{i\neq j} V_c(\mathbf{m}-\mathbf{n})\hat{n}_i\hat{n}_j + \sum_{\mathbf{q}} \omega_{\mathbf{q}}(d_{\mathbf{q}}^\dagger d_{\mathbf{q}} + 1/2). \end{aligned} \tag{1.32}$$

One can transform it further using the site representation also for phonons. Replacing the Bloch functions in the definition of $\gamma(\mathbf{q})$ by their Wannier series yields

$$\gamma(\mathbf{q}) = -\frac{1}{M^{1/2}\omega_{\mathbf{q}}^{3/2}} \sum_{\mathbf{n}} \mathrm{e}^{-\mathrm{i}\mathbf{q}\cdot\mathbf{n}} \mathbf{e}_{\mathbf{q}} \cdot \nabla_{\mathbf{n}} v(\mathbf{n}). \tag{1.33}$$

This result is obtained by neglecting the overlap integrals of the Wannier orbitals on different sites and by assuming that the single-ion potential $v(\mathbf{r})$ varies over the distance, which is much larger than the radius of orbits. Then using the displacement operators, one arrives at

$$H_{\text{e-ph}} = \sum_{\mathbf{m},\mathbf{n},s} \hat{n}_{\mathbf{m}s} \mathbf{u}_{\mathbf{n}} \cdot \nabla_{\mathbf{n}} v(\mathbf{m}-\mathbf{n}). \tag{1.34}$$

The site representation of $H_{\text{e-ph}}$ is particularly convenient for the interaction with dispersionless modes, when $\omega_{\mathbf{q}} = \omega_0$ and $\mathbf{e}_{\mathbf{q}} = \mathbf{e}$ are q-independent. Introducing the phonon site operators

$$d_{\mathbf{n}} = \frac{1}{\sqrt{N}} \sum_{\mathbf{q}} \mathrm{e}^{\mathrm{i}\mathbf{q}\cdot\mathbf{n}} d_{\mathbf{q}}, \tag{1.35}$$

one obtains in this case

$$\begin{aligned} \mathbf{u}_{\mathbf{n}} &= \frac{\mathbf{e}}{\sqrt{2M\omega_0}}(d_{\mathbf{n}} + d_{\mathbf{n}}^\dagger), \\ H_{\text{ph}} &= \omega_0 \sum_{\mathbf{n}} (d_{\mathbf{n}}^\dagger d_{\mathbf{n}} + 1/2), \end{aligned}$$

and

$$H_{\text{e-ph}} = \omega_0 \sum_{\mathbf{n},\mathbf{m},s} g(\mathbf{m}-\mathbf{n})(\mathbf{e}\cdot\mathbf{e}_{\mathbf{m}-\mathbf{n}})\hat{n}_{\mathbf{m}s}(d_{\mathbf{n}}^\dagger + d_{\mathbf{n}}), \tag{1.36}$$

where

$$g(\mathbf{m}) = \frac{1}{\omega_0\sqrt{2M\omega_0}} \frac{\mathrm{d}v(\mathbf{m})}{\mathrm{d}m}$$

is a dimensionless *force* acting between the electron on site $\mathbf{m}$ and the displacement of ion $\mathbf{n}$, and $\mathbf{e}_{\mathbf{m}-\mathbf{n}} \equiv (\mathbf{m}-\mathbf{n})/|\mathbf{m}-\mathbf{n}|$ is the unit vector in the direction from the electron $\mathbf{m}$ to the ion $\mathbf{n}$. The real-space representation is particularly convenient in parameterizing EPI in complex lattices. Atomic orbitals of an ion adiabatically follow its motion. Therefore, the electron does not interact with the displacement of the ion, which orbitals it occupies, i.e., $g(0) = 0$.

2

Continuum Polaron

If characteristic phonon frequencies are sufficiently low, the local deformation of ions, caused by electron itself, creates a potential well, which traps the electron even in a perfect crystal lattice. This *self-trapping* phenomenon was predicted by Landau [10] more than 70 years ago. It was studied in greater detail [11–15] in the effective mass approximation for the electron placed in a continuum polarizable (or deformable) medium, which leads to a so-called *large* or *continuum* polaron. Large-polaron wave functions and corresponding lattice distortions spread over many lattice sites. The self-trapping is never complete in the perfect lattice. Due to finite phonon frequencies, ion polarizations can follow polaron motion if the motion is sufficiently slow. Hence, large polarons with a low kinetic energy propagate through the lattice as free electrons but with an enhanced effective mass.

When the polaron binding energy E_{p} is larger than the half-bandwidth, D of the electron band, all states in the Bloch bands are "dressed" by phonons. In this strong-coupling regime, $\lambda = E_{\mathrm{p}}/D > 1$, the finite bandwidth becomes important, so the continuum approximation cannot be applied. In this case, the carriers are described as *small* or *lattice* polarons. The main features of small polarons were understood a long time ago [16–22]. The first identification of small polarons in solids was made for nonstoichiometric uranium dioxide in [23, 24]. Large and small polarons were discussed in a number of review papers and textbooks, for example [7, 11, 25–34].

In many models of EPI, the ground-state polaron energy is an analytical function of the coupling constant for any dimensionality of space [35–39]. There is no abrupt (nonanalytical) phase transition of the ground state as the electron–phonon coupling increases. It is instead a crossover from Bloch states of band electrons or large polarons propagating with almost bare mass in a rigid lattice to heavily dressed Bloch states of small polarons propagating at low temperatures with an exponentially enhanced effective mass. The ground-state wave function of any polaron is delocalized for any coupling strength. This result holds for both finite-site models and infinite-site models [38].

A.S. Alexandrov and J.T. Devreese, *Advances in Polaron Physics*, Springer Series in Solid-State Sciences 159,
DOI: 10.1007/978-3-642-01896-1_2, © Springer-Verlag Berlin Heidelberg 2010

2.1 Pekar's Polaron

To approach the multipolaron problem, let us briefly discuss a single electron interacting with the lattice deformation in the continuum approximation, as studied by Pekar [14]. In his model, a free electron interacts with the dielectric polarizable continuum, described by the static and high-frequency dielectric constants. This is the case for carriers interacting with optical phonons in ionic crystals under the condition that the size of the self-trapped state is large compared with the lattice constant, so that the lattice discreteness is irrelevant.

2.1.1 Ground State

Describing the ionic crystal as a polarizable dielectric continuum, one should keep in mind that only the ionic part of the total polarization contributes to the polaron state. The interaction of a carrier with valence electrons responsible for the optical properties is taken into account via the Hartree–Fock periodic potential and included in the band mass m. Therefore, only ion displacements contribute to the self-trapping. Following Pekar, we minimize the sum $E(\psi)$ of the electron kinetic energy and the potential energy due to the self-induced polarization field (here we follow [26]):

$$E(\psi) = \int d\mathbf{r} \left[\psi^*(\mathbf{r}) \left(-\frac{\nabla^2}{2m} \right) \psi(\mathbf{r}) - \mathbf{P}(\mathbf{r}) \cdot \mathbf{D}(\mathbf{r}) \right], \tag{2.1}$$

where

$$\mathbf{D}(\mathbf{r}) = e\nabla \int d\mathbf{r}' \frac{|\psi(\mathbf{r}')|^2}{|\mathbf{r} - \mathbf{r}'|} \tag{2.2}$$

is the electric field of the electron in the state with the wave function $\psi(\mathbf{r})$ and $\mathbf{P}$ is the ionic part of the lattice polarization. Minimizing $E(\psi)$ with respect to $\psi^*(\mathbf{r})$ at *fixed* $\mathbf{P}$ and $\int d\mathbf{r}|\psi(\mathbf{r})|^2 = 1$, one arrives at the equation of motion:

$$\left(-\frac{\nabla^2}{2m} - e \int d\mathbf{r}' \mathbf{P}(\mathbf{r}') \cdot \nabla' \frac{1}{|\mathbf{r}' - \mathbf{r}|} \right) \psi(\mathbf{r}) = E_0 \psi(\mathbf{r}), \tag{2.3}$$

where E_0 is the polaron ground-state energy. The ionic part of the total polarization is obtained using the definition of the susceptibilities χ_0 and χ, $\mathbf{P} = (\chi_0 - \chi)\mathbf{D}$. The dielectric susceptibilities χ_0 and χ are related to the static and high-frequency dielectric constants, respectively ($\chi_0 = (\epsilon_0 - 1)/4\pi\epsilon_0$, and $\chi = (\epsilon - 1)/4\pi\epsilon$), so that $\mathbf{P} = \mathbf{D}/4\pi\kappa$. Then the equation of motion becomes

$$\left(-\frac{\nabla^2}{2m} - \frac{e^2}{4\pi\kappa} \int d\mathbf{r}' \int d\mathbf{r}'' |\psi(\mathbf{r}'')|^2 \nabla' \frac{1}{|\mathbf{r}' - \mathbf{r}''|} \cdot \nabla' \frac{1}{|\mathbf{r}' - \mathbf{r}|} \right) \psi(\mathbf{r}) = E_0 \psi(\mathbf{r}). \tag{2.4}$$

Differentiating by parts with the use of $\nabla^2 r^{-1} = -4\pi\delta(\mathbf{r})$, one obtains

$$\left(-\frac{\nabla^2}{2m} - \frac{e^2}{\kappa}\int d\mathbf{r}' \frac{|\psi(\mathbf{r}')|^2}{|\mathbf{r}' - \mathbf{r}|}\right)\psi(\mathbf{r}) = E_0\psi(\mathbf{r}). \tag{2.5}$$

The solution of this nonlinear integrodifferential equation can be found using a variational minimization of the functional

$$J(\psi) = \frac{1}{2m}\int d\mathbf{r}|\nabla\psi(\mathbf{r})|^2 - \frac{1}{2ma_{\mathrm{B}}}\int d\mathbf{r}d\mathbf{r}' \frac{|\psi(\mathbf{r})|^2|\psi(\mathbf{r}')|^2}{|\mathbf{r}' - \mathbf{r}|}, \tag{2.6}$$

where $a_{\mathrm{B}} = \kappa/me^2$ is the effective Bohr radius. The simplest choice of the normalized trial function is $\psi(\mathbf{r}) = A\mathrm{e}^{-r/r_{\mathrm{p}}}$ with $A = (\pi r_{\mathrm{p}}^3)^{-1/2}$. Substituting the trial function into the functional yields $J(\psi) = T + \frac{1}{2}U$, where the kinetic energy is $T = 1/2mr_{\mathrm{p}}^2$, and the potential energy $U = -5/8ma_{\mathrm{B}}r_{\mathrm{p}}$. Minimizing the functional with respect to r_{p} yields the polaron radius, $r_{\mathrm{p}} = 16a_{\mathrm{B}}/5$, and the ground-state energy $E_0 = T + U$ as $E_0 = -0.146/ma_{\mathrm{B}}^2$. This can be compared with the ground-state energy of the hydrogen atom $-0.5m_{\mathrm{e}}e^4$, where m_{e} is the free electron mass. Their ratio is $0.3m/(m_{\mathrm{e}}\kappa^2)$. In typical polar solids $\kappa \gtrsim 4$, so that the continuum polaron binding energy is about 0.25 eV or less, if $m \simeq m_{\mathrm{e}}$. The potential energy in the ground state is $U = -4T = 4E_0/3$.

The lowest photon energy $\omega_{\min}$ to excite the polaron into the bare-electron band is $\omega_{\min} = |E_0|$. The ion configuration does not change in the photoexcitation process of the polaron. However, a lower activation energy W_{T} is necessary, if the self-trapped state disappears together with the polarization well due to thermal fluctuations, $W_{\mathrm{T}} = |E_0| - U_{\mathrm{d}}$, where U_{d} is the deformation energy. In ionic crystals,

$$U_{\mathrm{d}} = \frac{1}{2}\int d\mathbf{r}\mathbf{P}(\mathbf{r}) \cdot \mathbf{D}(\mathbf{r}), \tag{2.7}$$

which for the ground state is $U_{\mathrm{d}} = 2|E_0|/3$. Therefore, the thermal activation energy is $W_{\mathrm{T}} = |E_0|/3$. The ratio of four characteristic energies for the continuum Pekar's polaron is $W_{\mathrm{T}} : U_{\mathrm{d}} : \omega_{\min} : |U| = 1 : 2 : 3 : 4$ [40].

Using Pekar's choice,

$$\psi(r) = A(1 + r/r_{\mathrm{p}} + \beta r^2)\mathrm{e}^{-r/r_{\mathrm{p}}}, \tag{2.8}$$

one obtains $A = 0.12/r_{\mathrm{p}}^{3/2}$, $\beta = 0.45/r_{\mathrm{p}}^2$, the polaron radius $r_{\mathrm{p}} = 1.51a_{\mathrm{B}}$, and a better estimate for the ground-state energy, $E_0 = -0.164/ma_{\mathrm{B}}^2$ as compared to the result of the simplest exponential choice.

2.1.2 Effective Mass of Pekar's Polaron

The lattice polarization is responsible for the polaron mass enhancement [41]. Within the continuum approximation, the evolution of the lattice polarization

$\mathbf{P}(\mathbf{r},t)$ is described by a harmonic oscillator subjected to an external force $\sim\mathbf{D}/\kappa$:

$$\omega_0^{-2}\frac{\partial^2\mathbf{P}(\mathbf{r},t)}{\partial t^2}+\mathbf{P}(\mathbf{r},t)=\frac{\mathbf{D}(\mathbf{r},t)}{4\pi\kappa}, \tag{2.9}$$

where ω_0 is the optical phonon frequency. If during the characteristic time of the lattice relaxation $\simeq\omega_0^{-1}$ the polaron moves a distance much less than the polaron radius, the polarization practically follows the polaron motion. Hence for a slow motion with the velocity $v \ll \omega_0 a_\mathrm{B}$, the first term in (2.9) is a small perturbation, so that

$$\mathbf{P}(\mathbf{r},t)\approx\frac{1}{4\pi\kappa}\left(\mathbf{D}(\mathbf{r},t)-\omega_0^{-2}\frac{\partial^2\mathbf{D}(\mathbf{r},t)}{\partial t^2}\right). \tag{2.10}$$

The total energy of the crystal with an extra electron

$$E=E(\psi)+2\pi\kappa\int\mathrm{d}\mathbf{r}\left[\mathbf{P}(\mathbf{r},\mathbf{t})^2+\omega_0^{-2}\left(\frac{\partial\mathbf{P}(\mathbf{r},t)}{\partial t}\right)^2\right] \tag{2.11}$$

is determined in such a way that it gives (2.9) when it is minimized with respect to $\mathbf{P}$. We note that the first term of the lattice contribution to E is the deformation energy U_d, discussed in the previous section. The lattice part of the total energy depends on the polaron velocity and contributes to the effective mass. Replacing the static wave function $\psi(\mathbf{r})$ in all expressions for $\psi(\mathbf{r}-\mathbf{v}t)$ and neglecting a contribution to the total energy of higher orders than v^2, one obtains

$$E=E_0+U_\mathrm{d}+\frac{m^*v^2}{2}, \tag{2.12}$$

were

$$m^* = -\frac{1}{12\pi\omega_0^2\kappa}\int\mathrm{d}\mathbf{r}\mathbf{D}(\mathbf{r})\cdot\nabla^2\mathbf{D}(\mathbf{r}) \tag{2.13}$$

is the polaron mass. Using the equation $\nabla\cdot\mathbf{D}=-4\pi e|\psi(\mathbf{r})|^2$ yields

$$m^* = \frac{4\pi e^2}{3\omega_0^2\kappa}\int\mathrm{d}\,\mathbf{r}|\psi(\mathbf{r})|^4. \tag{2.14}$$

Calculating the integral with the trial function (2.8), one obtains

$$m^*\approx 0.02\alpha^4 m, \tag{2.15}$$

where α is the dimensionless constant, defined as $\alpha=(e^2/\kappa)\sqrt{m/2\omega_0}$.

Concluding the discussion of the Pekar's polarons, let us specify conditions of its existence. The polaron radius should be large compared with the lattice constant, $r_\mathrm{p}\gg a$ to justify the effective mass approximation for the electron. Hence, the value of the coupling constant α should not be very large, $\alpha \ll$

$(D/z\omega_0)^{1/2}$, where $D \simeq z/2ma^2$ is the bare half-bandwidth and z is the lattice coordination number. On the other hand, the classical approximation for the lattice polarization is justified if the number of phonons taking part in the polaron cloud is large. This number is of the order of U_d/ω_0. The total energy of the immobile polaron and the deformed lattice is $E = -0.109\alpha^2\omega_0$ and $U_\mathrm{d} = 0.218\alpha^2\omega_0$, respectively. Then α is bounded from below by the condition $U_\mathrm{d}/\omega_0 \gg 1$ as $\alpha^2 \gg 5$. The typical adiabatic ratio D/ω_0 is of the order of 10–100. In fact in many transition metal oxides with narrow bands and high optical phonon frequencies, this ratio is about 10 or even less, which makes the continuum strong-coupling polaron hard to be realized in oxides and related ionic compounds with light ions [26].

2.2 Fröhlich Large Polaron

2.2.1 Weak-Coupling Regime

Fröhlich [13] applied the second quantization form of the electron–lattice interaction [42] to describe the large polaron in the weak-coupling regime, $\alpha < 1$, where the quantum nature of lattice polarization becomes important:

$$H = -\frac{\nabla^2}{2m} + \sum_{\mathbf{q}} \left(V_{\mathbf{q}} d_{\mathbf{q}} \mathrm{e}^{\mathrm{i}\mathbf{q}\cdot\mathbf{r}} + \mathrm{H.c.}\right) + \sum_{\mathbf{q}} \omega_{\mathbf{q}} (d_{\mathbf{q}}^{\dagger} d_{\mathbf{q}} + 1/2). \tag{2.16}$$

The quantum states of the noninteracting electron and phonons are specified by the electron momentum $\mathbf{k}$ and the phonon occupation number $\langle d_{\mathbf{q}}^{\dagger} d_{\mathbf{q}} \rangle \equiv n_{\mathbf{q}} = 0, 1, 2, \ldots, \infty$. At zero temperature, the unperturbed state is the vacuum, $|0\rangle$, of phonons and the electron plane wave:

$$|\mathbf{k}, 0\rangle = \mathrm{e}^{\mathrm{i}\mathbf{k}\cdot\mathbf{r}} |0\rangle. \tag{2.17}$$

While the coupling is weak, one can apply the perturbation theory. The interaction couples the state (2.17) with the energy $k^2/2m$ and the state with the energy $(\mathbf{k} - \mathbf{q})^2/2m + \omega_0$ of a single phonon with momentum $\mathbf{q}$ and the electron with momentum $\mathbf{k} - \mathbf{q}$, $|\mathbf{k} - \mathbf{q}, 1_{\mathbf{q}}\rangle = \mathrm{e}^{\mathrm{i}(\mathbf{k}-\mathbf{q})\cdot\mathbf{r}} |1_{\mathbf{q}}\rangle$. The corresponding matrix element is $\langle \mathbf{k} - \mathbf{q}, 1_{\mathbf{q}} | H_{\text{e-ph}} | \mathbf{k}, 0 \rangle = V_{\mathbf{q}}^{*}$. There are no diagonal matrix elements of $H_{\text{e-ph}}$, so the second-order energy $\tilde{E}_{\mathbf{k}}$ is

$$\tilde{E}_{\mathbf{k}} = \frac{k^2}{2m} - \sum_{\mathbf{q}} \frac{|V_{\mathbf{q}}|^2}{(\mathbf{k} - \mathbf{q})^2/2m + \omega_0 - k^2/2m}. \tag{2.18}$$

There is no imaginary part of $\tilde{E}_{\mathbf{k}}$ for a slow electron with $k < q_\mathrm{p}$, where

$$q_\mathrm{p} = \min(m\omega_0/q + q/2) = (2m\omega_0)^{1/2}, \tag{2.19}$$

which means that the momentum is conserved. Evaluating the integrals one arrives at

$$\tilde{E}_{\mathbf{k}} = \frac{k^2}{2m} - \frac{\alpha\omega_0 q_{\mathrm{p}}}{k} \arcsin\left(\frac{k}{q_{\mathrm{p}}}\right), \tag{2.20}$$

which for a very slow motion $k \ll q_{\mathrm{p}}$ yields

$$\tilde{E}_{\mathbf{k}} \simeq -\alpha\omega_0 + \frac{k^2}{2m^*}. \tag{2.21}$$

Here, the first term is minus the polaron binding energy $-E_{\mathrm{p}}$. The effective mass of the polaron is enhanced as

$$m^* = \frac{m}{1-\alpha/6} \simeq m(1+\alpha/6) \tag{2.22}$$

due to a phonon "cloud" accompanying the slow polaron. The number of virtual phonons N_{ph} in the cloud is given by taking the expectation value of the phonon number operator, $N_{\mathrm{ph}} = \langle \sum_{\mathbf{q}} d^{\dagger}_{\mathbf{q}} d_{\mathbf{q}} \rangle$, where bra and ket refer to the perturbed state:

$$|\rangle = |0\rangle + \sum_{\mathbf{q}'} \frac{V^*_{\mathbf{q}'}}{k^2/2m - (\mathbf{k}-\mathbf{q}')^2/2m - \omega_0} |1_{\mathbf{q}'}\rangle. \tag{2.23}$$

For the polaron at rest ($\mathbf{k} = 0$), one obtains $N_{\mathrm{ph}} = \alpha/2$. Hence, the Fröhlich coupling constant, α, measures the cloud "thickness." One can also calculate the lattice charge density induced by the electron. The electrostatic potential $e\phi(\mathbf{r})$ is given by the average of the interaction term of the Hamiltonian

$$e\phi(\mathbf{r}) = \left\langle \sum_{\mathbf{q}} V_{\mathbf{q}} \mathrm{e}^{\mathrm{i}\mathbf{q}\cdot\mathbf{r}} d_{\mathbf{q}} + \mathrm{H.c.} \right\rangle, \tag{2.24}$$

and the charge density $\rho(\mathbf{r})$ is related to the electrostatic potential by Poisson's equation $\nabla\phi = -4\pi\rho$. Using these equations, one obtains

$$\rho(\mathbf{r}) = -\frac{1}{2\pi e} \sum_{\mathbf{q}} \frac{q^2 |V_{\mathbf{q}}|^2 \cos(\mathbf{q}\cdot\mathbf{r})}{\omega_0 + q^2/2m}, \tag{2.25}$$

which yields

$$\rho(\mathbf{r}) = -\frac{e q_{\mathrm{p}}^3}{4\pi\kappa} \frac{\mathrm{e}^{-q_{\mathrm{p}} r}}{q_{\mathrm{p}} r}. \tag{2.26}$$

The mean extension of the phonon cloud, which can be taken as the radius of the weak-coupling polaron, is $r_{\mathrm{p}} = q_{\mathrm{p}}^{-1}$, and the total induced charge is $Q = \int \mathrm{d}\mathbf{r} \rho(\mathbf{r}) = -e/\kappa$.

2.2.2 Lee–Low–Pines Transformation

One can put the Fröhlich result on a variational basis by applying the Lee–Low–Pines (LLP) canonical transformation [43], which removes the electron coordinate, followed by the displacement transformation [21,44,45]. The latter serves to account for that part of the lattice polarization which follows the electron instantaneously. The remaining part of the polarization field turns out to be small, if the coupling constant is not extremely large. In the opposite extreme limit, which is Pekar's strong-coupling regime discussed above, one can construct the perturbation theory by an expansion in descending powers of α [46–48]. Alternatively, one can apply Feynman's path-integral formalism to remove the phonon field at the expense of a noninstantaneous interaction of electron with itself (see (Sect. 2.3) and also [49]).

A canonical transformation can be written as

$$|\tilde{N}\rangle = \exp(S)|N\rangle, \tag{2.27}$$

where in our case $|N\rangle$ is a single-electron multiphonon wave function. The transformed eigenstate, $|\tilde{N}\rangle$ satisfies the Schrödinger equation, $\tilde{H}|\tilde{N} >= E|\tilde{N}\rangle$, with the transformed Hamiltonian

$$\tilde{H} = \exp(S)H\exp(-S). \tag{2.28}$$

If all operators are transformed according to (2.28), the physical averages remain unchanged. LLP transformation eliminating the electron coordinate in the Hamiltonian is defined as

$$S_{\mathrm{LLP}} = \mathrm{i}\sum_{\mathbf{q}}(\mathbf{q}\cdot\mathbf{r})d_{\mathbf{q}}^{\dagger}d_{\mathbf{q}}. \tag{2.29}$$

The transformed Hamiltonian is obtained as

$$\tilde{H} = \frac{1}{2m}\left(-\mathrm{i}\nabla - \sum_{\mathbf{q}}\mathbf{q}d_{\mathbf{q}}^{\dagger}d_{\mathbf{q}}\right)^2 + \sum_{\mathbf{q}}(V_{\mathbf{q}}d_{\mathbf{q}} + \mathrm{H.c.}) + \omega_0\sum_{\mathbf{q}}(d_{\mathbf{q}}^{\dagger}d_{\mathbf{q}} + 1/2). \tag{2.30}$$

The electron coordinate is absent in $\tilde{H}$. Hence, the eigenstates $|\tilde{N}\rangle$ are classified with the momentum $\mathbf{K}$, which is the conserving total momentum of the system, $|\tilde{N}\rangle = \mathrm{e}^{\mathrm{i}\mathbf{K}\cdot\mathbf{r}}|\tilde{N}_{\mathrm{ph}}\rangle$, where $|\tilde{N}_{\mathrm{ph}}\rangle$ is an eigenstate of phonons. The number of virtual phonons is not small in the intermediate-coupling regime. Therefore, one cannot apply the perturbation theory for $\tilde{H}$. However, one can remove the essential part of the interaction term in the Hamiltonian by the displacement canonical transformation:

$$S = \sum_{\mathbf{q}} f(\mathbf{q})d_{\mathbf{q}} - \mathrm{H.c.}, \tag{2.31}$$

where c-number $f(\mathbf{q})$ is determined by minimization of the ground-state energy. Assuming that the transformed ground state is the phonon vacuum $\mathrm{e}^S|\tilde{N}_{\mathrm{ph}}\rangle = |0\rangle$, one obtains the energy $E_{\mathbf{K}}$ as

$$E_{\mathbf{K}} = \frac{(1-\eta^2)K^2}{2m} - \frac{\alpha\omega_0 q_{\mathrm{p}}}{K(1-\eta)}\sin^{-1}\left(\frac{K(1-\eta)}{q_{\mathrm{p}}}\right), \tag{2.32}$$

where

$$\eta(1-\eta)^2 = \frac{\alpha q_{\mathrm{p}}^3}{2K^3}\left(\frac{(1-\eta)K}{\sqrt{q_{\mathrm{p}}^2-(1-\eta)^2K^2}} - \sin^{-1}\frac{(1-\eta)K}{q_{\mathrm{p}}}\right). \tag{2.33}$$

Only the term independent of K needs to be retained in η for a slow polaron with $K \ll q_{\mathrm{p}}$:

$$\eta = \frac{\alpha/6}{1+\alpha/6}. \tag{2.34}$$

Then, the energy up to the second order in K is given by

$$E_{\mathbf{K}} = -\alpha\omega_0 + \frac{K^2}{2m^*}, \tag{2.35}$$

where the polaron mass is $m^* = m(1+\alpha/6)$ as in (2.22). Lee, Low, and Pines evaluated also the corrections due to off-diagonal parts of the transformed Hamiltonian and found that they are small.

2.3 All-Coupling Fröhlich Polaron

2.3.1 Feynman Theory

Feynman developed a superior all-coupling continuum polaron theory using his path-integral formalism [12]. He got the idea to formulate the polaron problem into the Lagrangian form of quantum mechanics and then eliminate the field oscillators, "...in exact analogy to Q. E. D.... (resulting in)... a sum over all trajectories...." The resulting path integral (here limited to the ground-state properties) is of the form

$$\langle 0,\beta|0,0\rangle = \int \mathcal{D}\mathbf{r}(\tau)\mathrm{e}^{S}, \tag{2.36}$$

$$S = \exp\left[-\int_0^\beta \frac{\dot{\mathbf{r}}^2}{2}\mathrm{d}\tau + \frac{\alpha}{2^{3/2}}\int_0^\beta\int_0^\beta \frac{\mathrm{e}^{-|\tau-\sigma|}}{|\mathbf{r}(\tau)-\mathbf{r}(\sigma)|}\mathrm{d}\tau\mathrm{d}\sigma\right], \tag{2.37}$$

where $\beta = 1/T$. Equation (2.36) gives the *amplitude* that an electron found at a point in space at time zero will appear at the same point at the imaginary time β. The interaction term in the action function S may be interpreted as indicating that at "time" τ, the electron behaves as if it were in a potential:

$$\frac{\alpha}{2^{3/2}}\int_0^\beta \frac{\mathrm{e}^{-|\tau-\sigma|}}{|\mathbf{r}(\tau)-\mathbf{r}(\sigma)|}\mathrm{d}\sigma, \tag{2.38}$$

which results from the electrostatic interaction of the electron with the mean charge density of its "previous" positions, weighted with the function $e^{-|\tau-\sigma|}$. This path integral (2.36) with (2.37) has a great intuitive appeal: it shows the polaron problem as an equivalent one-particle problem in which the interaction, nonlocal in time or "retarded," occurs between the electron and itself.

Subsequently, Feynman introduced a variational principle for path integrals to study the polaron. He then simulated the interaction between the electron and the polarization modes by a harmonic interaction (with force constant k) between a hypothetical ("fictitious") particle with mass M and the electron. Within his model, the action function S (2.37) is imitated by a quadratic trial action (nonlocal in time):

$$S_0 = \exp\left[-\int_0^\beta \frac{\dot{\mathbf{r}}^2}{2}\mathrm{d}\tau + \frac{C}{2}\int_0^\beta\int_0^\beta [\mathbf{r}(\tau)-\mathbf{r}(\sigma)]^2\,\mathrm{e}^{-w|\tau-\sigma|}\mathrm{d}\tau\mathrm{d}\sigma\right], \quad (2.39)$$

where the interaction potential (2.38) is replaced by a parabolic potential:

$$\frac{C}{2}\int_0^\beta [\mathbf{r}(\tau)-\mathbf{r}(\sigma)]^2\,\mathrm{e}^{-w|\tau-\sigma|}\mathrm{d}\sigma \quad (2.40)$$

with the weight function $\mathrm{e}^{-w|\tau-\sigma|}$. The variational parameters C and w in (2.40) are adjusted to partly compensate for the error of exploiting the trial potential (2.40) instead of the true potential (2.38). Following the Feynman approach, an upper bound for the polaron ground-state energy can be written down as

$$E = E_0 - \lim_{\beta\to\infty}\frac{1}{\beta}\langle S - S_0\rangle_0, \quad (2.41)$$

where S is the exact action functional of the polaron problem, while S_0 is the trial action functional, which corresponds to the above model system, E_0 is the ground-state energy of the model system, and

$$\langle F\rangle_0 \equiv \frac{\int F\mathrm{e}^{S_0}\mathcal{D}\mathbf{r}\,(t)}{\int \mathrm{e}^{S_0}\mathcal{D}\mathbf{r}\,(t)}. \quad (2.42)$$

The parameters of the model system C and w are found from the variational condition that they provide a minimum to the upper bound for the ground-state energy of (2.41). (For the details of the calculation, see Sect. 2.5.) At nonzero temperatures, the best values of the model parameters can be determined from a variational principle for the free energy [50], see [51,52].

Equation (2.41) constitutes an upper bound for the polaron self-energy at all α, which in Fig. 2.1 is compared with the results of weak- and strong-coupling expansions. The weak-coupling expansions of Feynman for the ground-state energy and the effective mass of the polaron are

$$\frac{E_0}{\omega_0} = -\alpha - 0.0123\alpha^2 - 0.00064\alpha^3 - \cdots \quad (\alpha\to 0), \quad (2.43)$$

$$\frac{m^*}{m} = 1 + \frac{\alpha}{6} + 0.025\alpha^2 + \cdots \quad (\alpha\to 0). \quad (2.44)$$

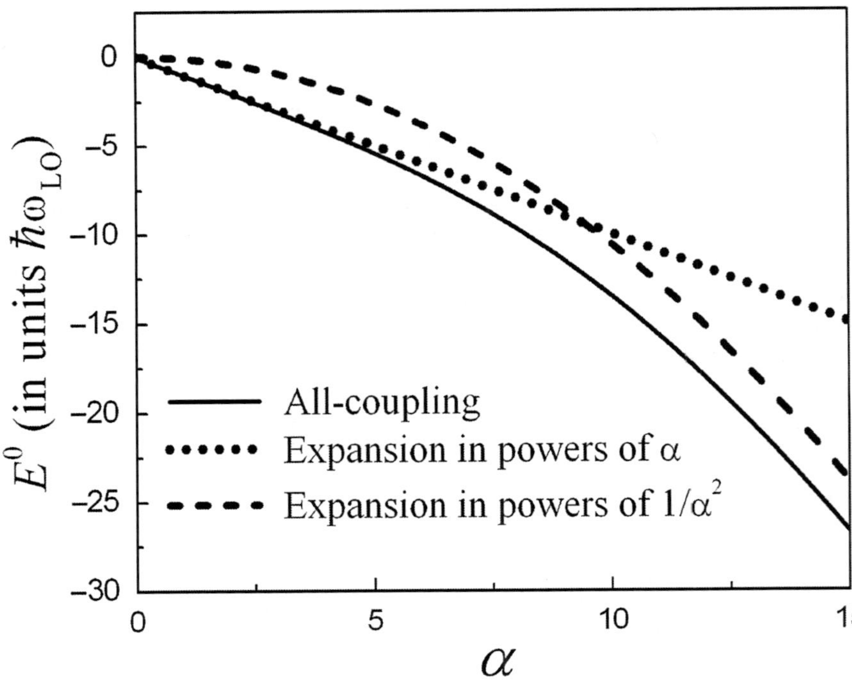

Fig. 2.1. Feynman polaron energy as a function of α: the all-coupling theory

In the strong-coupling limit, Feynman found for the ground-state energy

$$\frac{E_0}{\omega_0} \equiv \frac{E_{3\mathrm{D}}(\alpha)}{\omega_0} = -0.106\alpha^2 - 2.83 - \cdots \quad (\alpha \to \infty) \tag{2.45}$$

and for the polaron mass

$$\frac{m^*}{m} \equiv \frac{m^*_{3\mathrm{D}}(\alpha)}{m} = 0.0202\alpha^4 + \cdots \; (\alpha \to \infty). \tag{2.46}$$

Becker *et al.* [53], using a Monte Carlo calculation, derived the ground-state energy of a polaron as $E_0 = \lim_{\beta\to\infty} \Delta F$, where $\Delta F = F_\beta - F^0_\beta$ with F_β the free energy per polaron and $F^0_\beta = [3/(2\beta)] \ln(2\pi\beta)$ the free energy per electron. The value $\beta\omega_0 = 25$, used for the actual computation in [53], corresponds to $T/T_\mathrm{D} = 0.04$ ($T_\mathrm{D} = \hbar\omega_\mathrm{LO}/k_\mathrm{B}$; $\omega_\mathrm{LO} \equiv \omega_0$ is the longitudinal (LO) optical phonon frequency in conventional units). The authors of [53] actually calculated the *free energy* ΔF, rather than the polaron *ground-state energy*. To investigate the importance of temperature effects on ΔF, Peeters and Devreese [54] considered the polaron energy as obtained by Osaka [52], who generalized the Feynman polaron theory to nonzero temperatures:

$$\frac{\Delta F}{\omega_0} = \frac{3}{\beta} \ln\left(\frac{w}{v}\frac{\sinh\frac{\beta_0 v}{2}}{\sinh\frac{\beta_0 w}{2}}\right) - \frac{3}{4}\frac{v^2-w^2}{v}\left(\coth\frac{\beta_0 v}{2} - \frac{2}{\beta_0 v}\right)$$
$$- \frac{\alpha}{\sqrt{2\pi}}\left[1 + n(\omega_0)\right] \int_0^{\beta_0} \mathrm{d}u \frac{\mathrm{e}^{-u}}{\sqrt{D(u)}}, \tag{2.47}$$

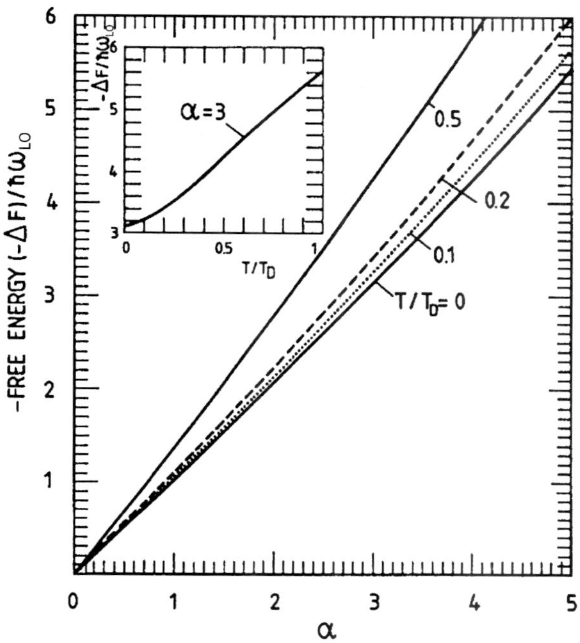

Fig. 2.2. Contribution of EPI to the free energy of the Feynman polaron as a function of the electron–phonon coupling constant α for different values of the lattice temperature. *Inset*: temperature dependence of the free energy for $\alpha = 3$ (from [54])

where $\beta_0 = \beta\omega_0$, $n(\omega) = 1/\left(\mathrm{e}^{\beta\omega} - 1\right)$, and

$$D(u) = \frac{w^2}{v^2}\frac{u}{2}\left(1 - \frac{u}{\beta_0}\right) + \frac{v^2 - w^2}{2v^3}\left(1 - \mathrm{e}^{-vu} - 4n(v)\sinh^2\frac{vu}{2}\right). \quad (2.48)$$

This result is variational, with variational parameters v and w, and gives an upper bound to the exact polaron free energy.

The results of a numerical-variational calculation of (2.47) are shown in Fig. 2.2, where the free energy $-\Delta F$ is plotted (in units of $\hbar\omega_{\mathrm{LO}}$) as a function of α for different values of the lattice temperature. As seen from Fig. 2.2, (a) $-\Delta F$ increases with increasing temperature and (b) the effect of temperature on ΔF increases with increasing α.

In Table 2.1, the Monte Carlo results [53], $(\Delta F)_{\mathrm{MC}}$, are compared with the free energy of the Feynman polaron, $(\Delta F)_{\mathrm{F}}$, calculated in [54]. The values for the free energy obtained analytically from the Feynman polaron model are *lower* than the published MC results for $\alpha \lesssim 2$ and $\alpha \geq 4$ (but lie within the 1% error of the Monte Carlo results). Since the Feynman result for the polaron free energy is an upper bound to the exact result, we conclude that for $\alpha \lesssim 2$ and $\alpha \geq 4$, the results of the Feynman model are closer to the exact result than the MC results of [53].

Table 2.1. Comparison between the free energy of the Feynman polaron theory, $-(\Delta F)_{\mathrm{F}}$, and the Monte Carlo results of [53], $-(\Delta F)_{\mathrm{MC}}$, for $T/T_{\mathrm{D}} = 0.04$

α	$-(\Delta F)_{\mathrm{F}}$	$-(\Delta F)_{\mathrm{MC}}$	Δ (%)
0.5	0.50860	0.505	0.71
1.0	1.02429	1.020	0.42
1.5	1.54776	1.545	0.18
2.0	2.07979	2.080	−0.010
2.5	2.62137	2.627	−0.21
3.0	3.17365	3.184	−0.32
3.5	3.73814	3.747	−0.24
4.0	4.31670	4.314	0.063

The relative difference is defined as $\Delta = 100 \times [(\Delta F)_{\mathrm{F}} - (\Delta F)_{\mathrm{MC}}]/(\Delta F)_{\mathrm{MC}}$ (from [54])

2.3.2 Diagrammatic Monte Carlo Algorithm

Mishchenko *et al.* [55] performed a study of the Fröhlich polaron on the basis of the diagrammatic quantum Monte Carlo (DQMC) method [56]. This method is based on the direct summation of Feynman diagrams for Green's functions in momentum space. The basic object of their investigation is the Matsubara Green's function of the polaron in the momentum ($\mathbf{k}$)-imaginary time (τ)-representation:

$$G(\mathbf{k},\tau) = \langle \mathrm{vac}|a_{\mathbf{k}}(\tau)a^{\dagger}_{\mathbf{k}}(0)|\mathrm{vac}\rangle, \quad \tau \geq 0, \tag{2.49}$$

$$a_{\mathbf{k}}(\tau) = \exp(H\tau)a_{\mathbf{k}}\exp(-H\tau). \tag{2.50}$$

In terms of a complete set $\{|\nu\rangle\}$ of eigenstates of the Fröhlich polaron Hamiltonian H, so that for a given $\mathbf{k}$ the relation $H|\nu(\mathbf{k})\rangle = E_{\nu}(\mathbf{k})|\nu((k))\rangle$ and $H|\mathrm{vac}\rangle = E_0|\mathrm{vac}\rangle$, the expansion of the Green's function (2.49)

$$G(\mathbf{k},\tau) = \sum_{\nu} |\langle \nu|a^{\dagger}_{\mathbf{k}}|\mathrm{vac}\rangle|^2 \exp\{-[E_{\nu}(\mathbf{k}) - E_0]\tau\} \tag{2.51}$$

follows straightforwardly. For the calculations discussed in what follows, $E_0 = 0$.

The spectral function (Lehmann function) $g_{\mathbf{k}}(\omega)$ is defined through the representation of the Green's function (2.51) in the form

$$G(\mathbf{k},\tau) = \int_0^{\infty} \mathrm{d}\Omega g_{\mathbf{k}}(\Omega), \tag{2.52}$$

$$g_{\mathbf{k}}(\Omega) = \sum_{\nu} \delta[\Omega - E_{\nu}(\mathbf{k})]|\langle \nu|a^{\dagger}_{\mathbf{k}}|\mathrm{vac}\rangle|^2. \tag{2.53}$$

This spectral function is normalized, $\int_0^{\infty} g_{\mathbf{k}}(\Omega)\mathrm{d}\Omega = 1$. It can be interpreted as the probability that a polaron has momentum $\mathbf{k}$ and energy Ω. The significance of the spectral function (2.53) is determined by the fact that it has poles (sharp peaks) at frequencies, which correspond to stable (metastable) quasiparticle states.

If, for a given $\mathbf{k}$, there is a stable state at energy $E(\mathbf{k})$, the spectral function takes the form

$$g_{\mathbf{k}}(\Omega) = Z_0^{(\mathbf{k})}\delta[\Omega - E(\mathbf{k})]\ldots, \tag{2.54}$$

where $Z_0^{(\mathbf{k})}$ is the weight of the bare-electron state. The energy $E_{\mathrm{gs}}(\mathbf{k})$ and the weight $Z_{0,\mathrm{gs}}^{(\mathbf{k})}$ for the polaron ground state can be extracted from the Green's function behavior at long times:

$$G(\mathbf{k}, \tau \gg \omega_0^{-1}) \to Z_0^{(\mathbf{k})} \exp[-E(\mathbf{k})\tau]. \tag{2.55}$$

Similarly to (2.49), the N-phonon Green's function is defined:

$$G_N(\mathbf{k}, \tau; \mathbf{q}_1, \ldots, \mathbf{q}_N) = \langle \mathrm{vac}|d_{\mathbf{q}_N}(\tau)\ldots d_{\mathbf{q}_1}(\tau)a_{\mathbf{p}}(\tau)a_{\mathbf{p}}^\dagger(0)d_{\mathbf{q}_1}^\dagger(0)\ldots d_{\mathbf{q}_N}^\dagger(0)|\mathrm{vac}\rangle,$$
$$\tau \geq 0,$$
$$\mathbf{p} = \mathbf{k} - \sum_{j-1}^{N} \mathbf{q}_j. \tag{2.56}$$

From the asymptotic properties of the Green's functions (2.56) at long times, the characteristics of the polaron ground state are found. In particular, the weight of the N-phonon state for the polaron ground state is given by

$$G_N(\mathbf{k}, \tau \gg \omega_0^{-1}; \mathbf{q}_1, \ldots, \mathbf{q}_N) \to Z_N^{(\mathbf{k})}(\mathbf{q}_1, \ldots, \mathbf{q}_N) \exp[-E(\mathbf{k})\tau]. \tag{2.57}$$

A standard diagrammatic expansion of the above-described Green's functions generates a series of Feynman diagrams. The following function is further introduced:

$$P(\mathbf{k}, \tau) = G(\mathbf{k}, \tau) + \sum_{N=1}^{\infty} \int \mathrm{d}\mathbf{q}_1 \ldots \mathrm{d}\mathbf{q}_N \tilde{G}_N(\mathbf{k}, \tau; \mathbf{q}_1, \ldots, \mathbf{q}_N), \tag{2.58}$$

where $\tilde{G}_N$ are irreducible N-phonon Green's functions (which do not contain disconnected phonon propagators). From (2.55), (2.57), and the completeness condition for the nondegenerate ground state

$$Z_0^{(\mathbf{k})} + \sum_{N=1}^{\infty} \int \mathrm{d}\mathbf{q}_1 \ldots \mathrm{d}\mathbf{q}_N Z_N^{(\mathbf{k})}(\mathbf{q}_1, \ldots, \mathbf{q}_N) = 1, \tag{2.59}$$

it follows that the polaron ground-state energy is determined by the asymptotic behavior of the function (2.58):

$$P(\mathbf{k}, \tau \gg \omega_0^{-1}) \to \exp[-E(\mathbf{k})\tau]. \tag{2.60}$$

The function $P(\mathbf{k},\tau)$ is an infinite series of integrals containing an ever-increasing number of integration variables. The essence of the DQMC method is to construct a process, which generates *continuum random variables* $(\mathbf{k},\tau)$ with a distribution function that coincides exactly with $P(\mathbf{k},\tau)$. Taking into account (2.58) and the diagrammatic rules, $P(\mathbf{k},\tau)$ is identified with the distribution function:

$$Q(\{y\}) = \sum_{m=0}^{\infty} \sum_{\xi_m} \int \mathrm{d}x_1 \ldots \mathrm{d}x_m F_m(\xi_m, \{y\}, x_1, \ldots, x_m), \tag{2.61}$$

where the external variables $\{y\}$ include $\mathbf{k}, \tau, \alpha$, and N, while the internal variables describe the topology of the diagram (labeled with ξ_m), times assigned to electron–phonon vertices and momenta of phonon propagators. The diagrammatic Monte Carlo process is a numeric procedure, which samples various diagrams in parameter space and collects statistics for $Q(\{y\})$ according to the Metropolis algorithm [57] in such a way that – when the process is repeated a large number of times – the result converges to the exact answer. The distribution function given by the convergent series (2.61) is simulated within the process of sequential stochastic generation of diagrams described by functions $F_m(\xi_m, \{y\}, x_1, \ldots, x_m)$. Further, using (2.52), the spectral function $g_{\mathbf{k}}(\omega)$ is obtained applying a stochastic optimization technique. We refer to the review [58] for further details on the DQMC and stochastic optimization, where information on the excited states of the polaron is also derived by the analytic continuation of the imaginary time Green's functions to real frequencies.

DQMC [55] (see Fig. 2.3) confirms that for $\alpha \gtrsim 1$, the bare-electron state in the polaron wave function is no longer the dominant contribution and perturbation theory is not adequate. The bare-electron weight Z_0^0 for the polaron ground state, as a function of the polaron coupling constant, rapidly vanishes for $\alpha \gtrsim 3$. It is suggested in [55] that in the interval $3 \lesssim \alpha \lesssim 10$, the polaron ground states smoothly transforms between weak- and strong-coupling limits.

In Sects. 2.4.3, 2.4.4, and 2.4.7, earlier analytical studies and results on Fröhlich polarons are compared with DQMC results. It would be beneficial, to have an independent numerical check of the DQMC results. The comparison of the DQMC results for the "low-energy" ($\Omega < 0$) part of the spectral density[1] for the Fröhlich polaron at $\alpha = 0.05$ [55] demonstrates perfect agreement with the perturbation theory result:

$$g_0(\Omega < 0) = \frac{\alpha}{2}\delta[\Omega + \alpha] \tag{2.62}$$

(m and ω_0 are set equal to unity). The DQMC results for the "high-energy" ($\Omega > 0$) part of the spectral density significantly differ from the perturbation

[1] It is worth of noticing that the spectral density is not identical to the optical absorption coefficient, which is discussed in Sects. 2.4.2–2.4.4.

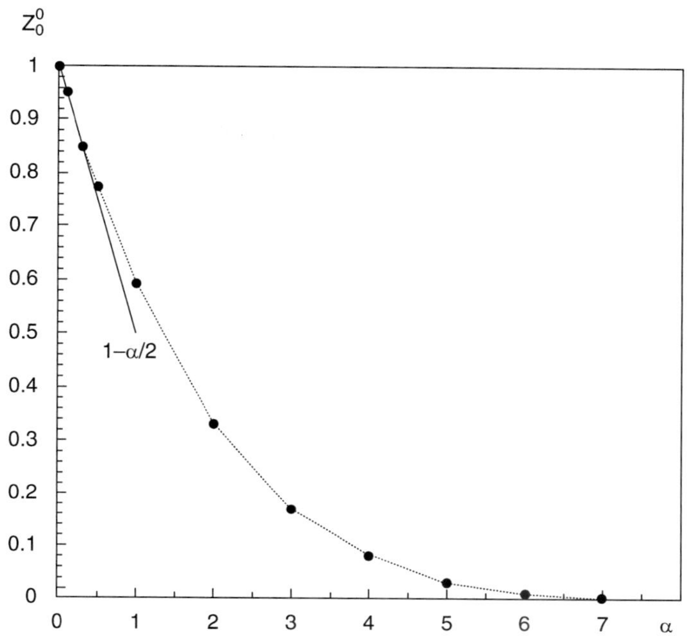

Fig. 2.3. The bare-electron factor Z_0^0 for the polaron ground state as a function of the coupling constant. *Filled circles* are the DQMC data (calculated to accuracy better than 10^{-4} in absolute values). The *dashed line* connects filled circles. The *solid line* is the perturbation theory result $Z_0^0 = 1 - \alpha/2$ (from [55])

theory curve. This is attributed to the fact that for the Fröhlich polaron, the perturbation theory expression for

$$g_0(\Omega > 0) = \frac{\alpha}{\pi} \frac{\theta(\Omega - 1)}{\Omega^2 \sqrt{\Omega - 1}} \tag{2.63}$$

diverges as $\Omega \to 1$ and, consequently, the perturbation approach is no longer adequate. The main difference between the DQMC spectrum of the Fröhlich polaron and the perturbation theory result is the broad peak in the spectral density at $\Omega \sim 3.5$. This peak appears for significantly larger values of the coupling constant and its weight grows with α (see Fig. 2.4). As shown in the inset, near the threshold, $\Omega = 1$, the spectral density behaves as $\sqrt{\Omega - 1}$.

Next to Fröhlich polarons, other polaron models have been extensively investigated using the Monte Carlo approach. In particular, Alexandrov [59] proposed a long-range *discrete* Fröhlich interaction as a model of the interaction between a hole and the oxygen ions in high-T_c superconductors. The essence of the model is that a charge carrier moves from site to site on a discrete lattice (or chain in 1D) and interacts with all the ions, which reside at the lattice sites. Numerically rigorous polaron characteristics (ground-state energy, number of phonons in the polaron cloud, effective mass, and isotope exponent) for such a lattice polaron, valid for arbitrary EPI, were

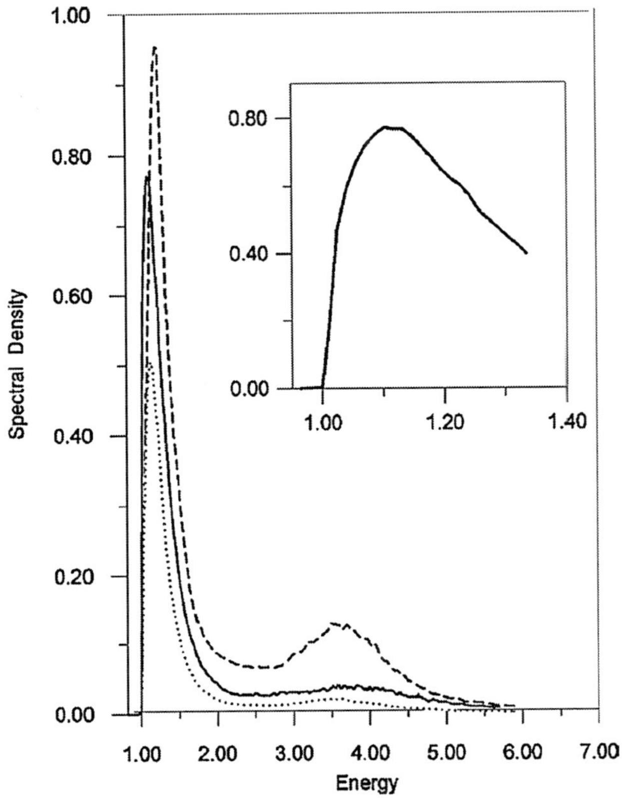

Fig. 2.4. The spectral density of the Fröhlich polaron for various values of the coupling constant: $\alpha = 0.5$ (*dotted line*), $\alpha = 1$ (*solid line*), and $\alpha = 2$ (*dashed line*) with energy counted from the ground-state energy of the polaron. The initial fragment of the spectral density for $\alpha = 1$ is shown in the *inset* (from [55])

obtained using a path-integral continuous-time quantum Monte Carlo technique (CTQMC) [37, 60, 61]. This study leads to a "mobile small Fröhlich" polaron (Sect. 3.4.3).

Over the years, the Feynman model for the Fröhlich polaron has remained the most successful approach to the problem. The analysis of an exactly solvable ("symmetrical") 1D-polaron model [62–64], Monte Carlo schemes [55, 65], and, recently, a unifying variational approach [66, 67] demonstrate the remarkable accuracy of Feynman's path-integral method. Using the variational wave functions, which combine both the Landau–Pekar and the Feynman approximations, Cataudella *et al.* [66] found in the $\alpha \to \infty$ limit

$$E_0 = \left(-0.108507\alpha^2 - 2.67\right)\omega_0. \tag{2.64}$$

which is slightly lower than the variational Feynman's estimate, (2.38), and at small values of α,

$$E_0 = -\alpha\omega_0 - 0.0123\alpha^2\omega_0, \quad \alpha \to 0, \tag{2.65}$$

i.e., the same result, as (2.43). In the latter limit, the correct result for the electron self-energy is

$$E = -\alpha\omega_0 - 0.0159\alpha^2\omega_0, \tag{2.66}$$

as found by Grosjean [68] and confirmed by Höhler and Mullensiefen [69] and Röseler [70].

Within the path-integral approach, Feynman *et al.* [71, 72] studied the mobility of polarons. Subsequently, the path-integral approach to the polaron problem was generalized and developed to become a powerful tool to study optical absorption and cyclotron resonance [73, 74].

2.4 Response of Continuum Polarons

2.4.1 Mobility

The mobility of large polarons was studied within various theoretical approaches. Fröhlich [13, 75] pointed out the typical behavior of the large-polaron mobility:

$$\mu \propto \exp(\omega_0\beta), \tag{2.67}$$

which is characteristic for weak coupling. Within the weak-coupling regime, the mobility of the polaron was then derived, e.g., using the Boltzmann equation in [76, 77] and starting from the LLP transformation in [78]. In [79] it was shown that for weak coupling,

$$\mu = \frac{e}{2\alpha}\left[\exp(\omega_0\beta) - 1\right]\left[1 - \frac{\alpha}{6} + O(\alpha^2)\right]. \tag{2.68}$$

A nonperturbative analysis was based on the Feynman polaron theory, where the mobility μ of the polaron using the path-integral formalism was derived by Feynman *et al.* [71] (FHIP) as a static limit, starting from a frequency-dependent impedance function. Details of the FHIP theory are given in [80]. An approximate expression for the impedance function of a Fröhlich polaron at all frequencies, temperatures, and coupling strengths was obtained in [71,80] within the path-integral technique. Assuming the crystal to be isotropic, an alternating electric field $\mathbf{E} = E_0\mathbf{e}_x \exp(\mathrm{i}\Omega t)$ induces a current in the x-direction:

$$j(\Omega) = \left[z(\Omega)\right]^{-1} E_0 \exp(\mathrm{i}\Omega t). \tag{2.69}$$

The complex function $z(\Omega)$ is called the impedance function. The electric field is considered sufficiently weak, so that linear-response theory can be applied. The frequency-depending mobility $\mu(\Omega)$ is defined as

$$\mu(\Omega) = \mathrm{Re}\left[z(\Omega)\right]^{-1}. \tag{2.70}$$

Taking the electric charge as unity, one arrives at $j = \langle \dot{x} \rangle$, where $\langle x \rangle$ is the expectation value of the electron displacement in the direction of the applied field: $\langle x \rangle = E/\mathrm{i}\Omega z(\Omega)$. In terms of time-dependent variables,

$$\langle x(\tau) \rangle = -\int_{-\infty}^{\tau} \mathrm{i}G(\tau - \sigma)E(\sigma)\mathrm{d}\sigma, \tag{2.71}$$

where $G(\tau)$ has the inverse Fourier transform:

$$G(\Omega) = \int_{-\infty}^{\infty} G(\tau) \exp(-\mathrm{i}\Omega\tau)\, \mathrm{d}\tau = [\Omega z(\Omega)]^{-1}. \tag{2.72}$$

The expected displacement at time τ is

$$\langle x(\tau) \rangle = \mathrm{Tr}\left[x U(\tau, a) \rho_a U'^{-1}(\tau, a) \right]. \tag{2.73}$$

Here, ρ_a is the density matrix of the system at some time a, long before the field is turned on, and

$$U(\tau, a) = T \exp\left\{ -\mathrm{i} \int_a^{\tau} [H_s - x_s E(s)]\, \mathrm{d}s \right\} \tag{2.74}$$

is the unitary operator of the development of a state in time with the complete Hamiltonian $H - xE$, where H is the Fröhlich polaron Hamiltonian, and T is the time-ordering operator [81]. Primed operators are ordered antichronologically:

$$U'^{-1}(\tau, a) = T' \exp\left\{ \mathrm{i} \int_a^{\tau} [H'_s - x'_s E'(s)]\, \mathrm{d}s \right\}. \tag{2.75}$$

$G(\tau - \sigma)$ can be represented as the second derivative:

$$G(\tau - \sigma) = \frac{1}{2} \left(\frac{\partial^2 g}{\partial \eta \partial \varepsilon} \right) \bigg|_{\varepsilon = \eta = 0} \tag{2.76}$$

of

$$g = \mathrm{Tr}\left[U(b, a) \rho_a U'^{-1}(b, a) \right], \quad b \to \infty, \quad a \to -\infty \tag{2.77}$$

with $E(s) = \varepsilon\delta(s - \sigma) + \eta\delta(s - \tau)$ and $E'(s) = \varepsilon\delta(s - \sigma) - \eta\delta(s - \tau)$. The initial state is chosen at a definite temperature T, $\rho_a \propto \exp(-\beta H)$. If the time a is sufficiently far in the past, FHIP assume that only the phonon subsystem was in thermal equilibrium at temperature T. The energy of the single electron and of the electron–phonon coupling are infinitesimal (of order $1/V$) with respect to that of the phonons. With this choice of the initial distribution, the phonon coordinates can be eliminated from (2.77), and the entire expression is reduced to a double path integral over the electron coordinates only:

$$g = \int \int \exp(\mathrm{i}\Phi)\, \mathcal{D}\mathbf{r}(t) \mathcal{D}\mathbf{r}'(t), \tag{2.78}$$

where (taking $m = 1$)

$$\Phi = \int_{-\infty}^{\infty} \left[\frac{\dot{\mathbf{r}}^2}{2} - \frac{\dot{\mathbf{r}}'^2}{2} \right] \mathrm{d}t - \int_{-\infty}^{\infty} \left[\mathbf{E}(t) \cdot \mathbf{r}(t) - \mathbf{E}'(t) \cdot \mathbf{r}'(t) \right] \mathrm{d}t \tag{2.79}$$

$$+ \frac{\mathrm{i}\alpha}{2^{3/2}} \int_{-\infty}^{\infty} \int_{-\infty}^{\infty} \left[\frac{\exp(-\mathrm{i}\,|t-s|) + 2P(\beta)\cos(t-s)}{|\mathbf{r}(t) - \mathbf{r}(s)|} \right.$$

$$+ \frac{\exp(+\mathrm{i}\,|t-s|) + 2P(\beta)\cos(t-s)}{|\mathbf{r}'(t) - \mathbf{r}'(s)|}$$

$$\left. - \frac{2\left[\exp(-\mathrm{i}\,(t-s)) + 2P(\beta)\cos(t-s)\right]}{|\mathbf{r}'(t) - \mathbf{r}(s)|} \right] \mathrm{d}t\mathrm{d}s,$$

where $P(\beta) = \left[\mathrm{e}^{\beta} - 1\right]^{-1}$. The double path integral in (2.78) is over paths which satisfy the boundary condition $\mathbf{r}(t) - \mathbf{r}'(t) = 0$ at times t approaching $\pm\infty$. Expression (2.78) with (2.79) is supposed to be exact [71]. Clearly to provide analytical solutions at all α presumably is impossible.

Following Feynman's idea to describe the ground-state energy properties of a polaron by introducing a parabolic "retarded" interaction of the electron with itself (see Sect. 2.3), it is assumed in [71] that the dynamical behavior of the polaron can be described approximately by replacing Φ of (2.78) by a parabolic (retarded) expression:

$$\Phi_0 = \int_{-\infty}^{\infty} \left[\frac{\dot{\mathbf{r}}^2}{2} - \frac{\dot{\mathbf{r}}'^2}{2} \right] \mathrm{d}t - \int_{-\infty}^{\infty} \left[\mathbf{E}(t) \cdot \mathbf{r}(t) - \mathbf{E}'(t) \cdot \mathbf{r}'(t) \right] \mathrm{d}t \tag{2.80}$$

$$- \frac{\mathrm{i}C}{2} \int_{-\infty}^{\infty} \int_{-\infty}^{\infty} \left\{ \left[\mathbf{r}(t) - \mathbf{r}(s)\right]^2 \left[\mathrm{e}^{-\mathrm{i}w|t-s|} + 2P(\beta w)\cos w(t-s)\right] \right.$$

$$+ \left[\mathbf{r}'(t) - \mathbf{r}'(s)\right]^2 \left[\mathrm{e}^{+\mathrm{i}w|t-s|} + 2P(\beta w)\cos(t-s)\right]$$

$$\left. - 2\left[\mathbf{r}'(t) - \mathbf{r}(s)\right]^2 \left[\mathrm{e}^{-\mathrm{i}w(t-s)} + 2P(\beta w)\cos w(t-s)\right] \right\} \mathrm{d}t\mathrm{d}s.$$

The parameters C and w are to be determined so as to approximate Φ as closely as possible. However, no variational principle is known for the mobility. At zero temperature, C and w are fixed at the values, which provide the best upper bound for the polaron ground-state energy (2.41). At finite temperatures ($T \neq 0$) the parameters C and w are determined from the variational principle for the polaron free energy [50]. This way of selection of the model parameters C and w is based on the supposition that "the comparison Lagrangian, which gives a good fit to the ground-state energy at zero temperature, will also give the dynamical behavior of the system" [71].

The analytical calculation of path integrals in (2.78) in [71] was performed to the first term in an expansion of $\exp\left[\mathrm{i}\,(\Phi - \Phi_0)\right]$:

$$g = \int\int \exp(\mathrm{i}\Phi_0) \exp[\mathrm{i}(\Phi - \Phi_0)] \mathcal{D}\mathbf{r}(t)\mathcal{D}\mathbf{r}'(t) \approx g_0 + g_1, \tag{2.81}$$

$$g_0 = \int\int \exp(\mathrm{i}\Phi_0) \mathcal{D}\mathbf{r}(t)\mathcal{D}\mathbf{r}'(t), \tag{2.82}$$

$$g_1 = i\int\int \exp(\mathrm{i}\Phi_0)(\Phi - \Phi_0) \mathcal{D}\mathbf{r}(t)\mathcal{D}\mathbf{r}'(t). \tag{2.83}$$

Using (2.72) and (2.76), one finds from (2.81)

$$G(\Omega) \approx G_0(\Omega) + G_1(\Omega). \tag{2.84}$$

The first term in the RHS of (2.84) is

$$G_0(\Omega) = \mathrm{i}Y_0(\Omega), \quad Y_0(\Omega) = -\frac{\Omega^2 - w^2}{(\Omega - \mathrm{i}\varepsilon)^2[(\Omega - \mathrm{i}\varepsilon)^2 - v^2]}, \quad \varepsilon \to +0 \tag{2.85}$$

with $v^2 = w^2 + 4C/w$. The second term in the RHS of (2.84) is

$$G_1(\Omega) = -\mathrm{i}Y_0^2(\Omega)\left[\chi(\Omega) - \frac{4C}{w}\frac{\Omega^2}{\Omega^2 - w^2}\right], \tag{2.86}$$

$$\chi(\Omega) = \int_0^\infty \left(1 - \mathrm{e}^{\mathrm{i}\Omega u}\right) \operatorname{Im} S(u) \mathrm{d}u, \tag{2.87}$$

$$S(u) = \frac{2\alpha}{3\sqrt{\pi}} \frac{\mathrm{e}^{\mathrm{i}u} + 2P(\beta)\cos u}{[D(u)]^{3/2}}, \tag{2.88}$$

$$D(u) = \frac{w^2}{v^2}\left\{\frac{v^2 - w^2}{w^2 v}\left[1 - \mathrm{e}^{\mathrm{i}vu} + 4P(\beta v)\sin^2\left(\frac{vu}{2}\right)\right] - \mathrm{i}u + \frac{u^2}{\beta}\right\}. \tag{2.89}$$

From (2.72) and (2.84), the impedance results in the form

$$\Omega z(\Omega) \approx [G_0(\Omega) + G_1(\Omega)]^{-1}. \tag{2.90}$$

Feynman *et al.* [71] suggested to use this expression expanded to first order in $G_1(\Omega)$:

$$\Omega z(\Omega) \approx [G_0(\Omega)]^{-1} - [G_0(\Omega)]^{-2} G_1(\Omega). \tag{2.91}$$

Substitution of (2.85) and (2.86) into (2.91) leads to the final expression [71] for the impedance function of the Fröhlich polaron:

$$\Omega z(\Omega) = \mathrm{i}\left[\Omega^2 - \chi(\Omega)\right]. \tag{2.92}$$

An alternative derivation of the impedance function of the Fröhlich polaron, based on simple operator techniques, was presented in [82]. The FHIP result was worked out in [83] in detail to get a physical insight into the scattering processes incorporated in the FHIP approximation. For sufficiently low temperature, the FHIP polaron mobility takes the form [71]

$$\mu = \left(\frac{w}{v}\right)^3 \frac{3e}{4m\omega_0^2\alpha\beta} e^{\omega_0\beta} \exp\{(v^2 - w^2)/w^2 v\}, \tag{2.93}$$

where v and w are variational functions of α obtained from the Feynman polaron model.

Using the Boltzmann equation for the Feynman polaron model, Kadanoff [84] found the mobility, which for low temperatures can be represented as follows:

$$\mu = \left(\frac{w}{v}\right)^3 \frac{e}{2m\omega_0\alpha} e^{\omega_0\beta} \exp\{(v^2 - w^2)/w^2 v\}, \tag{2.94}$$

The weak-coupling perturbation expansion of the low-temperature polaron mobility as found using the Green's function technique [79] has confirmed that the mobility derived from the Boltzmann equation is asymptotically exact for weak coupling ($\alpha \ll 1$) and at low temperatures ($T \ll \omega_0$). As first noticed in [84], the mobility of (2.93) differs by the factor of $3/(2\beta\omega_0)$ from that derived using the polaron Boltzmann equation as given by (2.94).[2] As follows from this comparison, the result of [71] is not valid when $T \to 0$. As emphasized in [71] and later confirmed, in particular, in [83, 85] the above discrepancy can be attributed to an interchange of two limits in calculating the impedance. In FHIP, for weak electron–phonon coupling, one takes $\lim_{\Omega\to 0}\lim_{\alpha\to 0}$, whereas $\lim_{\alpha\to 0}\lim_{\Omega\to 0}$ should be calculated. It turns out that for the asymptotically correct result, the mobility at low temperatures is mainly limited by the absorption of phonons, while in the theory of FHIP it is the emission of phonons which gives the dominant contribution as T goes to zero [85].

The analysis based on the Boltzmann equation takes into account the phonon emission processes whenever the energy of the polaron is above the emission threshold. The independent-collision model, which underlies the Boltzmann equation approach, however, fails in the "strong-coupling regime" of the Fröhlich polaron, when the thermal mean free path becomes less than the de Broglie wavelength; in this case, the Boltzmann equation cannot be expected to be adequate [71, 86].

Experimental work on alkali halides and silver halides indicates that the mobility obtained from (2.93) describes the experimental results accurately (see [87, 88] and references therein). Measurements of mobility as a function of temperature for photoexcited electrons in cubic n-type $Bi_{12}SiO_{20}$ (Fig. 2.5) are explained well in terms of large polarons within the Feynman

[2] In the asymptotic limit of weak electron–phonon coupling and low temperature, the FHIP polaron mobility of (2.93) differs by the same factor of $3/(2\beta\omega_0)$ from the earlier result [76–78], which, as pointed out in [71] and in later publications [79, 83, 84], is correct for $\beta \gg 1$.

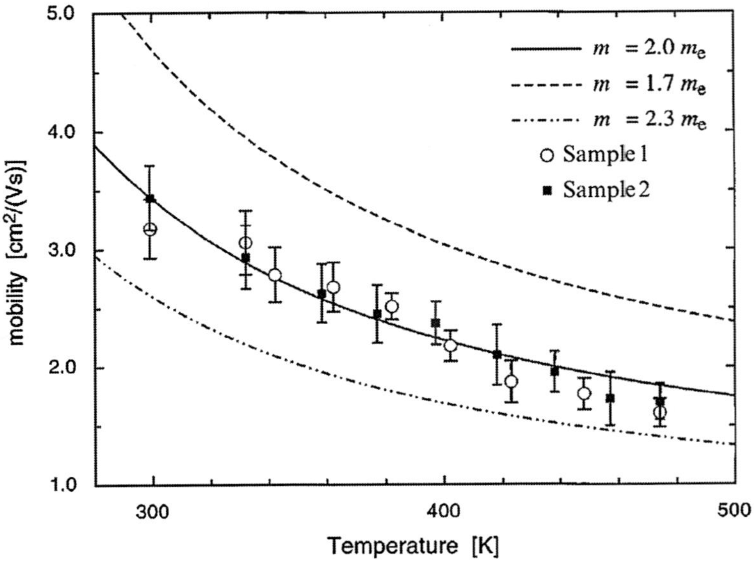

Fig. 2.5. Comparison of experimental and theoretical mobilities for cubic n-type $Bi_{12}SiO_{20}$. *Open circles and squares* represent experimental band mobility data for photoexcited electrons in two n-$Bi_{12}SiO_{20}$ samples. The three theoretical curves are drawn for $m/m_e = 1.7$, 2.0, and 2.3 using the Feynman polaron model. The calculated mobility at $T = 300\,\mathrm{K}$ corresponds to the measured room-temperature mobility value of $3.4\,\mathrm{cm}^2\,\mathrm{Vs}^{-1}$ when setting $m = 2.01 m_e$ (after [89])

approach [86]. The experimental findings on electron transport in crystalline TiO_2 (rutile phase) probed by THz time-domain spectroscopy were quantitatively interpreted within the Feynman model [90]. One of the reasons for the agreement between theory based on (2.93) and experiment is that in the path-integral approximation to the polaron mobility, a Maxwellian distribution for the *electron velocities* is assumed, when applying the adiabatic switching of the Fröhlich interaction. Although such a distribution is not inherent in the Fröhlich interaction, its incorporation tends to favor agreement with experiment because other mechanisms (interaction with acoustic phonons, etc.) cause a Gaussian distribution.

2.4.2 Optical Absorption at Weak Coupling

At zero temperature and in the weak-coupling limit, the optical absorption of a Fröhlich polaron is due to the elementary polaron scattering process with the absorption of incoming photon and emission of an outgoing phonon. In the weak-coupling limit ($\alpha \ll 1$), the polaron absorption coefficient for a many-polaron gas was first obtained by Gurevich *et al.* [91]. Their optical absorption coefficient is equivalent to a particular case of [92] with the dynamic structure factor $S(\mathbf{q},\Omega)$ corresponding to the Hartree–Fock approximation. In [92], the optical absorption coefficient of a many-polaron gas was shown to be given, to order α, by

$$\mathrm{Re}[\sigma(\Omega)] = n_{\mathrm{p}}\, e^2 \frac{2}{3}\alpha \frac{1}{2\pi\Omega^3} \int_0^\infty \mathrm{d}q q^2 S(\mathbf{q},\Omega-\omega_0), \tag{2.95}$$

where n_{p} is the density of charge carriers.

In the zero-temperature limit, starting from the Kubo formula the optical conductivity of a *single* Fröhlich polaron can be represented in the form

$$\begin{aligned}\sigma(\Omega) = \mathrm{i}\frac{e^2}{m\,(\Omega+\mathrm{i}\delta)} + \frac{e^2}{m^2}\frac{1}{(\Omega+\mathrm{i}\delta)^3}\int_0^\infty \mathrm{e}^{-\delta t}\left(\mathrm{e}^{\mathrm{i}\Omega t}-1\right) \\ \times \sum_{\mathbf{q},\mathbf{q}'} q_x q_x' \left\langle \Psi_0 | [\hat{B}_{\mathbf{q}}(t), \hat{B}_{-\mathbf{q}'}(0)] | \Psi_0 \right\rangle \mathrm{d}t,\end{aligned} \tag{2.96}$$

where $\delta = +0$, $\hat{B}_{\mathbf{q}}(t) = [V_{\mathbf{q}} d_{\mathbf{q}}(t) + V_{-\mathbf{q}}^* d_{-\mathbf{q}}^\dagger(t)]\mathrm{e}^{\mathrm{i}\mathbf{q}\cdot\mathbf{r}(t)}$, and $|\Psi_0\rangle$ is the ground-state wave function of the electron–phonon system. Within the weak-coupling approximation, the following analytic expression for the real part of the polaron optical conductivity results from (2.96):

$$\mathrm{Re}\,\sigma(\Omega) = \frac{\pi e^2}{2m^*}\delta(\Omega) + \frac{2e^2}{3m}\frac{\omega_0\alpha}{\Omega^3}\sqrt{\Omega-\omega_0}\Theta(\Omega-\omega_0), \tag{2.97}$$

where $\Theta(x) = 1$ if $x > 0$ and zero otherwise. The spectrum of the real part of the polaron optical conductivity (2.97) is represented in Fig. 2.6.

According to (2.97), the absorption coefficient of light with frequency $\Omega > 0$ by free polarons for $\alpha \to 0$ takes the form

$$\Gamma(\Omega) = \frac{1}{\epsilon_{\mathrm{v}} n}\frac{2n_{\mathrm{p}}\, e^2\alpha\omega_0^2}{3m\Omega^3}(\Omega/\omega_0 - 1)^{1/2}\Theta(\Omega-\omega_0), \tag{2.98}$$

where ϵ_{v} is the dielectric permittivity of the vacuum, and n is the refractive index of the medium. A simple derivation [95] using a canonical transformation method gives the absorption coefficient of free polarons, which coincides with the result (2.98). The step function in (2.98) reflects the fact that at zero temperature, the absorption of light accompanied by the emission of a phonon can occur only if the energy of the incident photon is larger than that of a phonon ($\Omega > \omega_0$). In the weak-coupling limit, according to (2.98), the absorption spectrum consists of a "one-phonon line." At nonzero temperature, the absorption of a photon can be accompanied not only by emission, but also by absorption of one or more phonons. Similarity between the temperature dependence of several features of the experimental infrared absorption spectra in high-T_{c} superconductors and the temperature dependence predicted for the optical absorption of a single Fröhlich polaron [73, 93] has been revealed in [96].

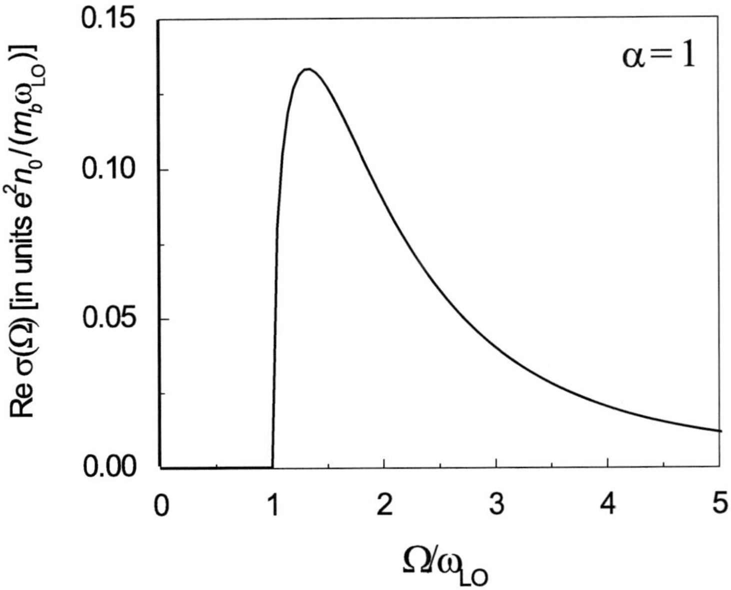

Fig. 2.6. Polaron optical conductivity for $\alpha = 1$ in the weak-coupling approximation, according to [93]. A δ-like central peak (at $\Omega = 0$) is schematically shown by a *vertical line* (reprinted with permission from [94]. © 2006, Società Italiana di Fisica)

Experimentally, this one-phonon line has been observed for free polarons in the infrared absorption spectra of CdO-films [97], which is a weakly polar material with $\alpha \approx 0.74$. The polaron absorption band is observed in the spectral region between 6 and 20 µm (above the LO phonon frequency). The difference between the one-polaron theoretical absorption and experiment in the wavelength region where polaron absorption dominates the spectrum has been explained as due to many-polaron effects [92].

2.4.3 Optical Absorption at Strong Coupling

The structure of the large-polaron excitation spectrum constituted a central question at the early stages of the development of polaron theory. The exactly solvable polaron model of [62] was used to demonstrate the existence of the so-called "relaxed excited states" of large polarons [63]. In [62], an exactly solvable ("symmetric") 1D-polaron model was introduced and analyzed. The further study of this model was performed in [63, 64]. The model consists of an electron interacting with two oscillators possessing opposite wave vectors: $\mathbf{q}$ and $-\mathbf{q}$. The parity operator, which changes $d_{\mathbf{q}}$ and $d_{-\mathbf{q}}$ (and also $d_{\mathbf{q}}^{\dagger}$ and $d_{-\mathbf{q}}^{\dagger}$), commutes with the Hamiltonian of the system. Hence, the polaron states are classified into even and odd states with eigenvalues of the parity operator +1 and −1, respectively. For the lowest even and odd states, the phonon distribution functions W_N are plotted in Fig. 2.7, upper panel, for some values of the effective coupling constant λ of this "symmetric" model. The value of the parameter $\varkappa = [q^2/m\omega_0]^{1/2}$ for these graphs was taken to be 1, while the total polaron momentum $\mathbf{P} = 0$. In the weak-coupling case ($\lambda \approx 0.6$), W_N is a decaying function of N. When increasing λ, W_N acquires a maximum, e.g., at $N = 8$ for the lowest even state at $\lambda = 5.0625$. The phonon distribution function W_N has the same character for the lowest even and the lowest odd states at all values of the number of virtual phonons in the ground state (as distinct from the higher states). This led to the conclusion that the lowest odd state is an internal excited state of the polaron.

In [55], the structure of the Fröhlich polaron cloud was investigated using the DQMC method. Contributions of N-phonon states to the polaron ground state were calculated as a function of N for a few values of the coupling constant α (see Fig. 2.7, lower panel). As follows from Fig. 2.7, the evolution from the weak-coupling case ($\alpha = 1$) into the strong-coupling regime ($\alpha = 17$) was studied. The evolution of the shape and the scale of the distribution of the N-phonon states with increasing α as derived for a Fröhlich polaron within DQMC method [55] is *in notable agreement* with the results obtained within the "symmetric" 1D-polaron model [62–64].

The insight gained as a result of those investigations concerning the structure of the excited polaron states was subsequently used to develop a theory of the optical absorption spectra of polarons. The first work was limited to the strong-coupling limit [98], where the impact of the internal degrees of freedom of polarons on their optical properties was revealed. The optical absorption of light by free Fröhlich polarons was treated in [98] using the polaron states obtained within the adiabatic strong-coupling approximation. It was argued in [98] that for sufficiently large α ($\alpha \gtrsim 3$), the (first) relaxed excited state (RES) of a polaron is a relatively stable state, which gives rise to a "resonance" in the polaron optical absorption spectrum. The following scenario of a transition, which leads to a "zero-phonon" peak in the absorption by a strong-coupling polaron, was then suggested. If the frequency of the incoming photon is equal to $\Omega_{\mathrm{RES}} = 0.065\alpha^2\omega_0$, the electron jumps from the ground state (which, at large coupling, is well characterized by "s"-symmetry for the electron) to an excited state ("2p"), while the lattice polarization in the final state is adapted to the "2p" electronic state of the polaron. In [98], considering the decay of the RES with emission of one real phonon, it is argued that the "zero-phonon" peak can be described using the Wigner–Weisskopf formula which is valid when the linewidth of that peak is much smaller than ω_0.

For photon energies larger than $\Omega_{\mathrm{RES}} + \omega_{\mathrm{LO}}$, a transition of the polaron toward the *first scattering state*, belonging to the RES, becomes possible. The final state of the optical absorption process then consists of a polaron in its lowest RES plus a free phonon. A "one-phonon sideband" then appears in the polaron absorption spectrum. This process is called *one-phonon sideband*

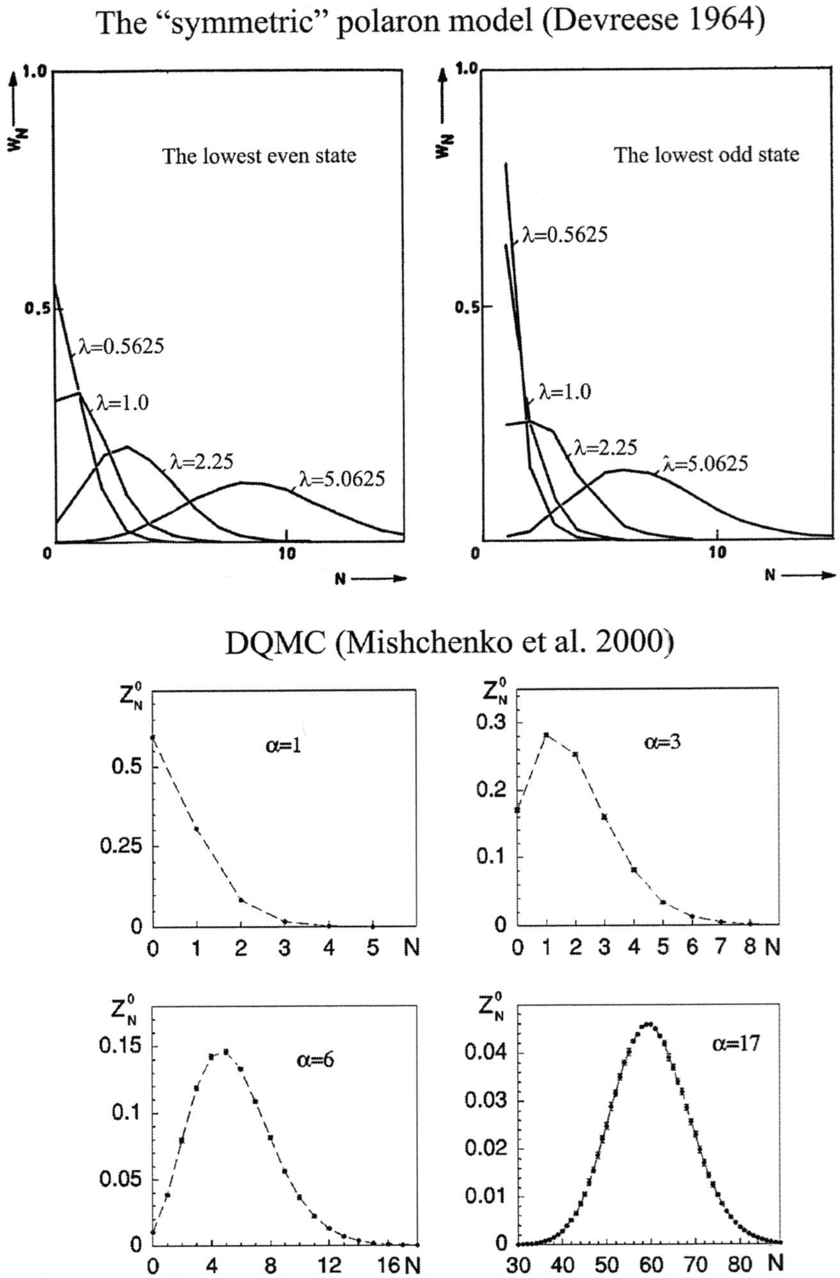

Fig. 2.7. *Upper panel*: the phonon distribution functions W_N in the "symmetric" polaron model for various values of the effective coupling constant λ at $\varkappa = 1, \mathbf{P} = 0$ (from [62]). *Lower panel*: distribution of multiphonon states in the polaron cloud within DQMC method for various values of α (from [55]) (reprinted with permission from [94]. © 2006, Società Italiana di Fisica)

absorption. The one-, two-, ..., K-, ... phonon sidebands of the zero-phonon peak give rise to a broad structure in the absorption spectrum. It turns out that the *first moment* of the phonon sidebands corresponds to the Franck–Condon (FC) frequency $\Omega_{\mathrm{FC}} = 0.141\alpha^2\omega_0$.

To summarize, following [98], the polaron optical absorption spectrum at strong coupling is characterized by the following features (at $T = 0$):

(a) An absorption peak ("zero-phonon line") appears, which corresponds to a transition from the ground state to the first RES at Ω_{RES}.
(b) For $\Omega > \Omega_{\mathrm{RES}} + \omega_0$, a phonon sideband structure arises. This sideband structure peaks around Ω_{FC}. Even when the zero-phonon line becomes weak, and most oscillator strength is in the LO-phonon sidebands, the zero-phonon line continues to determine the onset of the phonon sideband structure.

The basic qualitative strong-coupling behavior predicted in [98], namely, zero-phonon (RES) line with a broader sideband at the high-frequency side, was confirmed by later studies, as discussed below.

2.4.4 Optical Absorption at Arbitrary Coupling

The optical absorption of the Fröhlich polaron was calculated in 1972 [73,93] ("DSG") for the Feynman polaron model using path integrals. Until recently, DSG combined with Kartheuser *et al.* [98] constituted the basic picture for the optical absorption of the Fröhlich polaron. Peeters and Devreese [82] red-erived the DSG result using the memory function formalism (MFF). The DSG approach is successful at weak electron–phonon coupling and is able to identify the excitations at intermediate electron–phonon coupling ($3 \lesssim \alpha \lesssim 6$). In the strong-coupling limit, DSG still gives an accurate first moment for the polaron optical absorption but does not reproduce the broad phonon sideband structure (cf. [98,99]). A comparison of the DSG results with the optical conductivity spectra given by recently developed "approximation-free" numerical [100] and approximate analytical [67, 101] approaches was carried out recently [101], see also the review articles [58, 66].

The polaron absorption coefficient $\Gamma(\Omega)$ of light with frequency Ω at arbitrary coupling was first derived in [73]. It was represented in the form

$$\Gamma(\Omega) = -\frac{1}{n\epsilon_{\mathrm{v}}}\frac{e^2}{m}\frac{\operatorname{Im}\Sigma(\Omega)}{[\Omega - \operatorname{Re}\Sigma(\Omega)]^2 + [\operatorname{Im}\Sigma(\Omega)]^2}. \tag{2.99}$$

This general expression was the starting point for a derivation of the theoretical optical absorption spectrum of a single Fröhlich polaron at *all electron–phonon coupling strengths* by Devreese *et al.* [73]. $\Sigma(\Omega)$ is the so-called "memory function," which contains the dynamics of the polaron and depends on Ω, α, temperature, and applied external fields. The key contribution of [73] was to introduce $\Gamma(\Omega)$ in the form (2.99) and to calculate $\operatorname{Re}\Sigma(\Omega)$, which is essentially a (technically not trivial) Kramers–Kronig transform of the more simple function $\mathrm{Im}\,\Sigma(\Omega)$. Only the function $\mathrm{Im}\,\Sigma(\Omega)$ had been derived for the Feynman polaron [71] to study the polaron mobility μ using the impedance function (2.70): $\mu^{-1} = \lim_{\Omega\to 0} (\mathrm{Im}\Sigma(\Omega)/\Omega)$.

The basic nature of the Fröhlich polaron excitations was clearly revealed through this polaron optical absorption given by (2.99). It was demonstrated [73] that the FC states for Fröhlich polarons are nothing else but a superposition of phonon sidebands and a relatively large value of the electron–phonon coupling strength ($\alpha > 5.9$) is needed to stabilize the relaxed excited state of the polaron. It was, further, revealed that at weaker coupling only "scattering states" of the polaron play a significant role in the optical absorption [73, 102].

In the weak-coupling limit, the absorption spectrum (2.99) of the polaron is determined by the absorption of radiation energy, which is re-emitted in the form of LO phonons. As α increases between approximately 3 and 6, a resonance with increasing stability appears in the optical absorption of the Fröhlich polaron of [73] (see Fig. 2.8). The RES peak in the optical absorption spectrum also has a phonon sideband structure, whose average transition frequency can be related to a FC-type transition. Furthermore, at zero temperature, the optical absorption spectrum of one polaron exhibits also a zero-frequency "central peak" [$\propto \delta(\Omega)$]. For nonzero temperature, this "central peak" smears out and gives rise to an "anomalous" Drude-type low-frequency component of the optical absorption spectrum.

For $\alpha > 6.5$, the polaron optical absorption gradually develops the structure *qualitatively* proposed in [98]: a broad LO-phonon sideband structure appears with the zero-phonon ("RES") transition as onset. Devreese et al. [73]

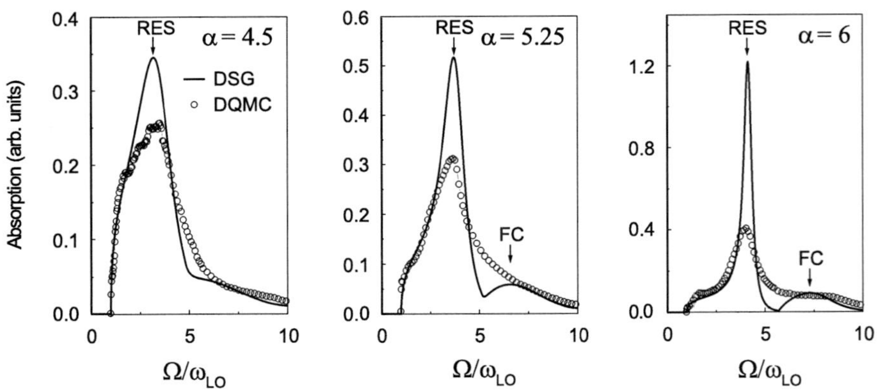

Fig. 2.8. Optical absorption spectrum of a Fröhlich polaron for $\alpha = 4.5$, $\alpha = 5.25$, and $\alpha = 6$ (after [73]) (DSG). The RES peak is very intense compared with the FC peak. The δ-like central peaks (at $\Omega = 0$) are schematically shown by *vertical lines*. The DQMC results of [100] are shown with *open circles*

does not predict the *broad* LO-phonon sidebands at large coupling constant, although it still gives an accurate first Stieltjes moment of the optical absorption spectrum.

In Fig. 2.8, the main peak of the polaron optical absorption for $\alpha = 5.25$ at $\Omega = 3.71\omega_0$ is interpreted as due to transitions to a RES. The "shoulder" at the low-frequency side of the main peak is attributed as mainly due to one-phonon transitions to polaron "scattering states." The broad structure centered at about $\Omega = 6.6\omega_0$ is interpreted as a FC band (composed of LO-phonon sidebands). As seen from Fig. 2.8, when increasing the electron–phonon coupling constant to $\alpha = 6$, the RES peak at $\Omega = 4.14\omega_0$ stabilizes. Up to $\alpha = 6$, the DQMC results of [100] reproduce the main features of the optical absorption spectrum of a Fröhlich polaron as found in [73].

Based on Devreese *et al.* [73], it was argued that it is rather Holstein polarons that determine the optical properties of the charge carriers in oxides like $SrTiO_3$, $BaTiO_3$ [103], while Fröhlich weak-coupling polarons could be identified, e.g., in CdO [95]. The Fröhlich coupling constants of polar semiconductors and ionic crystals are generally too small to allow for a static "RES." In [104], the RES peaks of [73] were invoked to explain the optical absorption spectrum of $Pr_2NiO_{4.22}$. The RES-like resonances in $\Gamma(\Omega)$ (2.99) due to the zeros of $\Omega - \mathrm{Re}\,\Sigma(\Omega)$ can effectively be displaced to smaller polaron coupling by applying an external magnetic field B, in which case the contribution for what is formally a "RES-type resonance" arises at $\Omega - \omega_c - \mathrm{Re}\,\Sigma(\Omega) = 0$ ($\omega_c = eB/m$ is the cyclotron frequency). Resonances in the magnetoabsorption governed by this contribution have been clearly observed and analyzed in many solids and structures, see, e.g., [11, 74, 105–108] and references therein.

Evidence for the polaron character of charge carriers in AgBr and AgCl was obtained through high-precision cyclotron resonance experiments in external magnetic fields up to 16 T (see Fig. 2.9). The all-coupling magnetoabsorption calculated in [74] leads to the best quantitative agreement between theory and experiment for AgBr and AgCl. This quantitative interpretation of the cyclotron resonance experiment in AgBr and AgCl [106] by the theory of [74] provided one of the most convincing and clearest demonstrations of Fröhlich polaron features in solids. The energy spectra of polaronic systems such as shallow donors ("bound polarons"), e.g., the D_0 and D^- centers, constitute the most complete and detailed polaron spectroscopy realized in the literature (see, e.g., Fig. 2.10).

The numerical calculations of the optical conductivity for the Fröhlich polaron performed within the DQMC method [100] confirm the analytical results derived in [73] for $\alpha \lesssim 3$. In the intermediate-coupling regime $3 < \alpha < 6$, the low-energy behavior and the position of the RES peak in the optical conductivity spectrum of [100] follow closely the prediction of [73]. There are some minor quantitative differences between the two approaches in the intermediate-coupling regime: in [100], the dominant ("RES") peak is less intense in the Monte Carlo numerical simulations and the second ("FC") peak develops less prominently. The following qualitative differences exist between

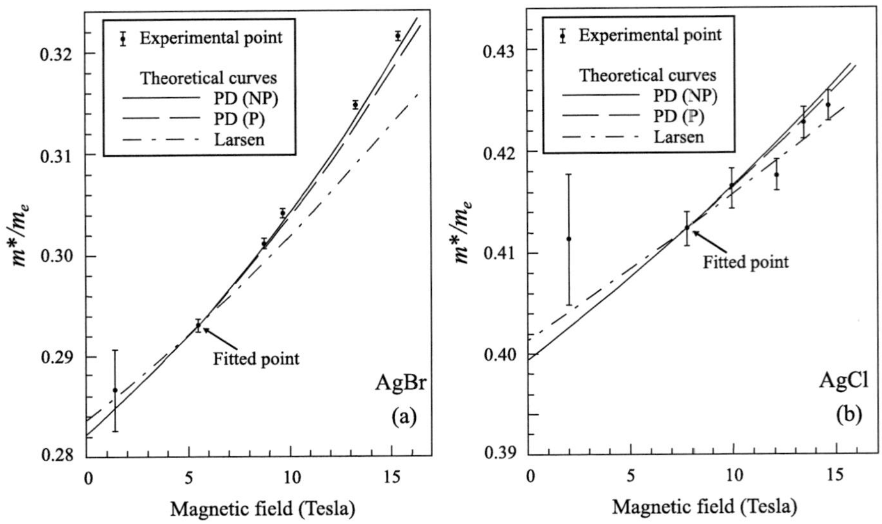

Fig. 2.9. The polaron cyclotron mass in AgBr (**a**) and in AgCl (**b**): comparison of experiment and theory (Larsen: [109]; PD: [74]); *P* parabolic band, *NP* corrections of a two-band Kane model. In each case, the band mass was adjusted to fit the experimental point at 525 GHz (reprinted with permission from [106]. © 1987 by the American Physical Society)

the two approaches: in [100], the dominant peak broadens for $\alpha \gtrsim 6$ and the second peak does not develop, but gives rise to a flat shoulder in the optical conductivity spectrum at $\alpha \approx 6$. As α increases beyond $\alpha \approx 6$, the DSG results for the OC do not produce the *broad* phonon sideband spectrum of the RES transition that was *qualitatively* predicted in [98] and obtained with DQMC.

Figure 2.11 shows that already for $\alpha = 1$ noticeable differences arise between $\mathrm{Re}\,\sigma(\Omega)$ calculated with perturbation theory to $O(\alpha)$, respectively, $O(\alpha^2)$, and DSG or DQMC. Remarkably, the DQMC results for $\alpha = 1$ seem to show a somewhat more pronounced two-phonon scattering contribution than the perturbation theory result to $O(\alpha^2)$. This point deserves further analysis.

An instructive comparison between the positions of the main peak in the optical absorption spectra of Fröhlich polarons obtained within the DSG and DQMC approaches has been performed recently [112]. In Fig. 2.12, the frequency of the main peak in the OC spectra calculated within the DSG approach [73] is plotted together with that given by DQMC [100, 101]. As seen from the figure, the main peak positions, obtained within DSG, are in good agreement with the results of DQMC for all considered values of α. At large α, the positions of the main peak in the DSG spectra are remarkably close to those given by DQMC. The difference between the DSG and DQMC

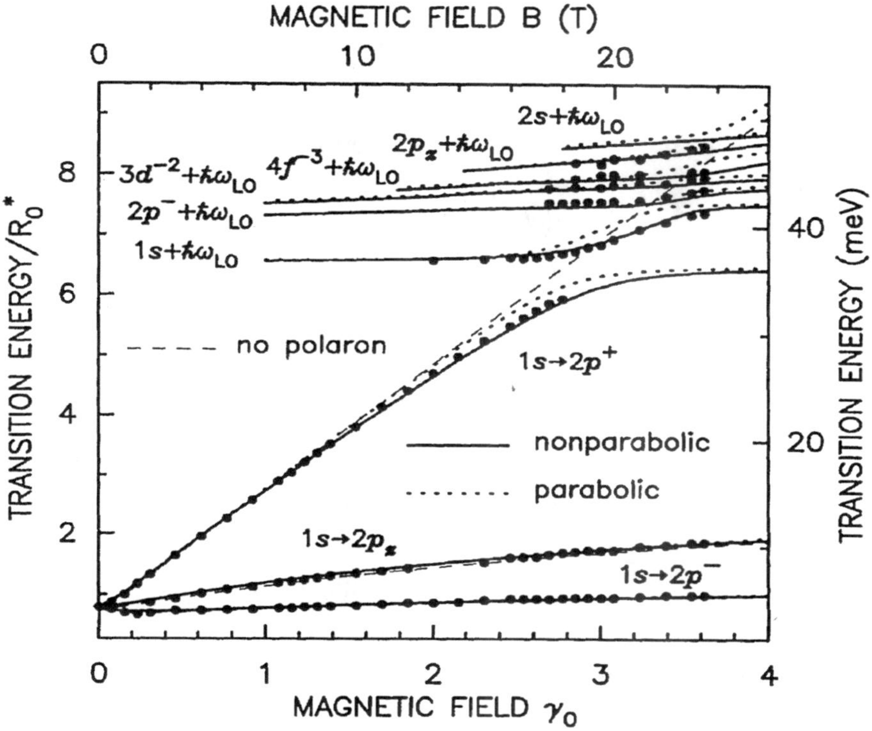

Fig. 2.10. The $1s \to 2p^{\pm}, 2p_z$ transition energies as a function of a magnetic field for a donor in GaAs. The authors of [108] compare their theoretical results for the following cases: (**a**) without the effect of polaron and band nonparabolicity (*thin dashed curves*); (**b**) with polaron correction (*dotted curves*); (**c**) including the effects of polaron and band nonparabolicity (*solid curves*) to the experimental data of [110] (*solid dots*). Here $\gamma_0 = 0.15413\,(\mathrm{T})$ is the dimensionless unit of the magnetic field (reprinted with permission from [108]. © 1993 by the American Physical Society)

results is relatively larger at $\alpha = 8$ and for $\alpha = 9.5$, but even for those values of the coupling constant the agreement is rather good.

It is suggested that the RES peak at $\alpha \approx 6$ in the DSG treatment, as α increases, gradually transforms onto a FC peak. As stated above and in [73], DSG predicts a much too narrow FC peak in the strong-coupling limit, but still at the "correct" frequency. The DSG spectrum also satisfies the *zero* and *first* moment sum rules at all α as discussed below.

2.4.5 Main Peak Line and Strong-Coupling Expansion

To describe the OC main peak linewidth at intermediate electron–phonon coupling, the DSG approach was modified [101] to include additional dissipation processes, the strength of which is fixed by an exact sum rule [66].

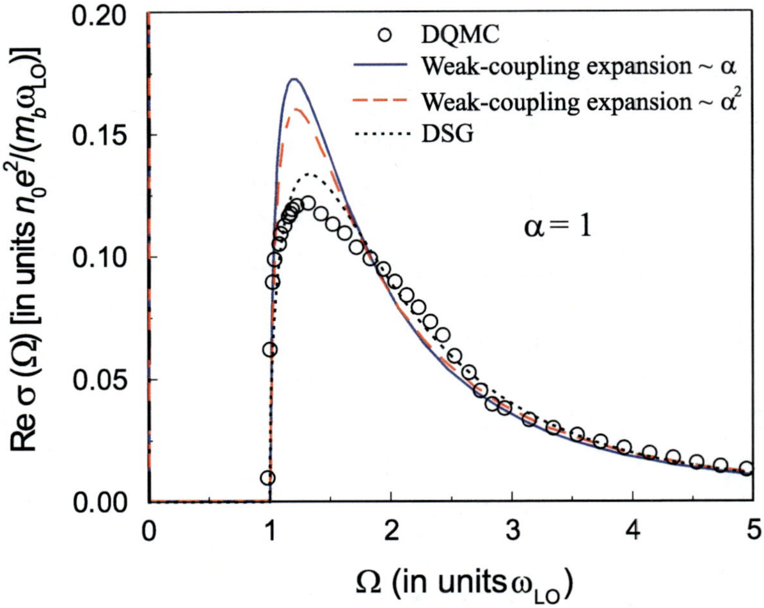

Fig. 2.11. One-polaron optical conductivity $\mathrm{Re}\,\sigma(\Omega)$ for $\alpha = 1$ calculated within the DQMC approach [100] (*open circles*), derived using the expansion in powers of α up to α [95] (*solid line*) up to α^2 [111] (*dashed line*) and within the DSG approach [73] (*dotted line*). A δ-like central peak (at $\Omega = 0$) is schematically shown by a vertical line

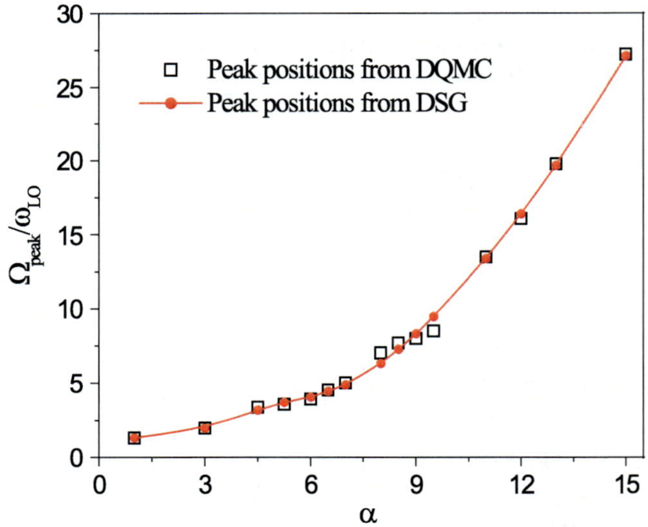

Fig. 2.12. Main peak positions from DQMC optical conductivity spectra of Fröhlich polarons [101] compared to those of the analytical DSG approach [73] (from [112])

Within the memory function formalism (MFF) [113, 114], the interaction of the charge carriers with the free phonon oscillations is expressed in terms of the electron density–density correlation function:

$$\chi(\mathbf{q},t) = -\mathrm{i}\Theta(t)\left\langle \exp\left[\mathrm{i}\mathbf{q}\cdot\mathbf{r}(t)\right]\exp\left[-\mathrm{i}\mathbf{q}\cdot\mathbf{r}(0)\right]\right\rangle, \tag{2.100}$$

which is evaluated in a direct way [82] using the Feynman polaron model, where the electron is coupled via a harmonic force to a fictitious particle that simulates the phonon degrees of freedom. Within this procedure, the electron density–density correlation function takes the form:

$$\chi_m(\mathbf{q},t) = -\mathrm{i}\Theta(t)\exp\left[-\mathrm{i}q^2t/2M\right]\exp\left[-q^2R(1-\mathrm{e}^{-\mathrm{i}vt})/2M\right], \tag{2.101}$$

where $R = (M-1)/v$, M (the total mass of electron and fictitious particle), and v are determined variationally within the path-integral approach [12]. The associated spectral function $A_m(\mathbf{q},\omega) = -2\mathrm{Im}\,\chi_m(\mathbf{q},\omega)$ is a series of δ-functions centered at $q^2/2M + nv$ (n is integer). Here, $q^2/2M$ represents the energy of the center of mass of electron and fictitious particle, and v is the energy gap between the levels of the relative motion. To include dissipation [101], a finite lifetime was introduced for the states of the relative motion, which can be considered as the result of the residual EPI not included into the Feynman variational model. To this end, in $\chi_m(\mathbf{q},t)$ the factor $\exp[-ivt]$ was replaced with $(1+\mathrm{i}t/\tau)^{-v\tau}$ which leads to the replacement of δ-functions by Gamma functions with mean value and variance given, respectively, by $q^2/2M + nv$ and nv/τ. The parameter of dissipation τ is determined by the third sum rule for $A(\mathbf{q},\omega)$, which is additional to the first two sum rules that are already satisfied in the DSG model without damping. As expected, τ turns out to be of the order of ω_0^{-1}. If broadening of the oscillator levels is neglected, the DSG results [73, 82] are recovered.

Starting from the Kubo formula, the strong-coupling polaron optical conductivity can be evaluated using the strong-coupling expansion (SCE) [66, 101]. In [112, 115], SCE has been extended. To apply the extended SCE for the polaron OC, a scaling transformation of the coordinates and moments of the electron–phonon system is made following Allcock [46], $\mathbf{r} = \alpha^{-1}\mathbf{x}$, $\mathbf{p} = -\mathrm{i}\alpha\partial/\partial\mathbf{x}$, and $\mathbf{q} = \alpha\tilde{\mathbf{q}}$. This transformation allows us to see explicitly the order of magnitude of different terms in the Hamiltonian. Expressed in terms of the new variables, the Hamiltonian can be written as a sum of two terms, which are of different orders in powers of α, $H = H_1 + H_2$, where $H_1 \sim \alpha^2$ is the leading term and $H_2 \sim \alpha^{-2}$ is the kinetic energy of the vibrating ions. The next step is the Born–Oppenheimer approximation [46], which neglects the nonadiabatic transitions between different polaron levels in calculating the dipole–dipole correlation function of the Kubo formula [112, 115].

Figure 2.13 shows the polaron OC spectra for different values of α calculated numerically using the extended SCE with different approximations. The OC spectra calculated within the extended SCE approach, taking into account both the Jahn–Teller effect and the corrections of order α^0, are

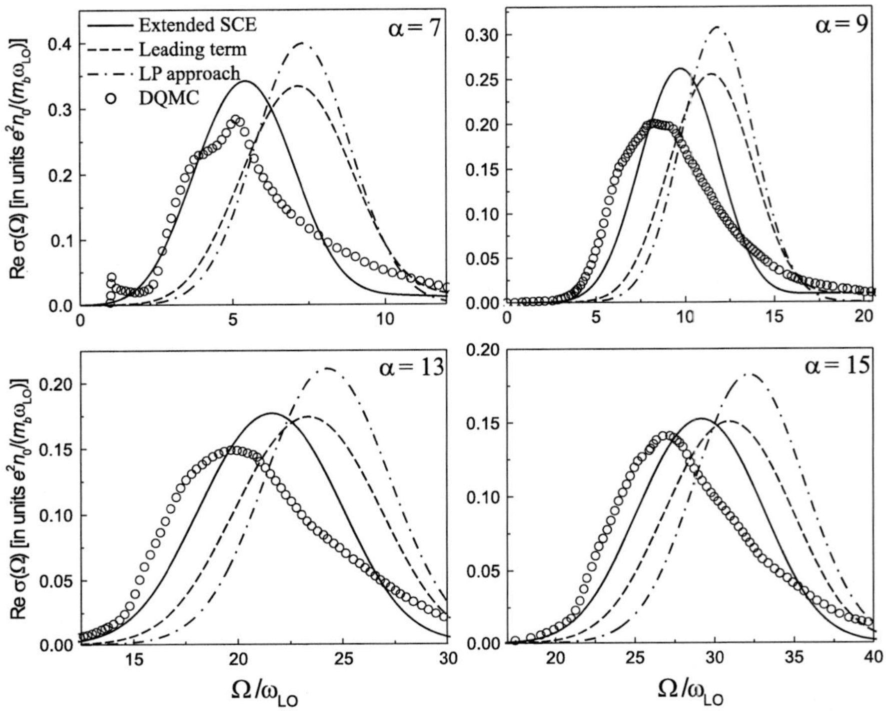

Fig. 2.13. The polaron OC calculated within the extended SCE taking into account corrections of order α^0 (*solid curve*), the OC calculated within the leading-term strong-coupling approximation (*dashed curve*), with the leading term of the Landau–Pekar (LP) adiabatic approximation (*dash-dotted curve*), and the numerical DQMC data (*open circles*) for $\alpha = 7$, 9, 13, and 15 (from [112])

shown by the solid curves. The OC obtained with the leading-term strong-coupling approximation taking into account the Jahn–Teller effect and with the leading term of the Landau–Pekar adiabatic approximation are plotted as dashed and dash-dotted curves, respectively. The full circles show the DQMC data [100, 101].

The polaron OC band of Fig. 2.13 obtained within the extended SCE generalizes the Gaussian-like polaron OC band (as given, e.g., by equation (3) of [101]), thanks to (1) the use of the numerically exact strong-coupling polaron wave functions [116] and (2) the incorporation of both static and dynamic Jahn–Teller effects. The polaron OC broad structure calculated within the extended SCE consists of a series of LO-phonon sidebands and provides a realization – with all LO-phonons involved for a given α – of the strong-coupling scheme proposed in [98].

As seen from Fig. 2.13, the polaron OC spectra calculated within the asymptotically exact strong-coupling approach are shifted toward lower frequencies as compared with the OC spectra calculated within the LP approximation. This shift is due to the use of the numerically exact (in the strong-coupling limit) energy levels and wave functions of the internal excited polaron states, as well as the numerically exact self-consistent adiabatic polaron potential. Furthermore, the inclusion of the corrections of order α^0 leads to a shift of the OC spectra to lower frequencies with respect to the OC spectra calculated within the leading-term approximation. The value of this shift $\Delta\Omega_{n,0}/\omega_{LO} \approx -1.8$ obtained in the present calculation is close to the LP value $\Delta\Omega_{n,0}^{(LP)}/\omega_0 = 4\ln 2 - 1 \approx 1.7726$ (cf. [50, 117]). The distinction between the OC spectra calculated with and without the Jahn–Teller effect is very small.

Starting from $\alpha \approx 9$ toward larger values of α, the agreement between the extended SCE polaron OC spectra and the numerical DQMC data becomes gradually better, consistent with the fact that the extended SCE for the polaron OC is asymptotically correct in the strong-coupling limit. The results of the extended SCE are qualitatively consistent with the interpretation advanced in [98]. In [98], only the 1-LO-phonon sideband was taken into account, while in [99] 2-LO-phonon emission was included. The extended SCE carries on the program started in [98]. The spectra in Fig. 2.13, in the strong-coupling approximation, consist of LO-phonon sidebands to the RES (which itself has negligible oscillator strength in this limit, similar to the optical absorption for some color centers in alkali halides). These LO-phonon sidebands form a broad FC-structure.

2.4.6 Comparison Between Optical Conductivity Spectra Obtained by Different Methods

A comparison between the optical conductivity spectra obtained with the DQMC method, extended MFF, SCE, and DSG for different values of α is shown in Figs. 2.14 and 2.15, taken from [101]. The key results of the comparison are the following.

First, as expected, in the weak-coupling regime, both the extended MFF with phonon broadening and DSG [73] are in very good agreement with the

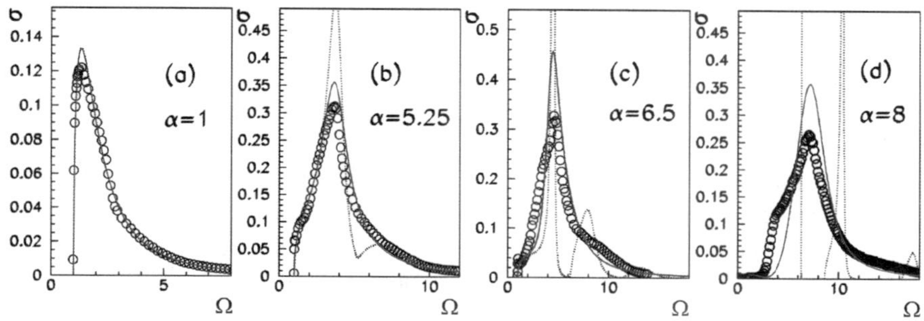

Fig. 2.14. Comparison of the optical conductivity calculated with the DQMC method (*circles*), extended MFF (*solid line*), and DSG [73,82] (*dotted line*), for four different values of α. The *arrow* indicates the lower-frequency feature in the DQMC data (reprinted with permission from [101]. © 2006 by the American Physical Society)

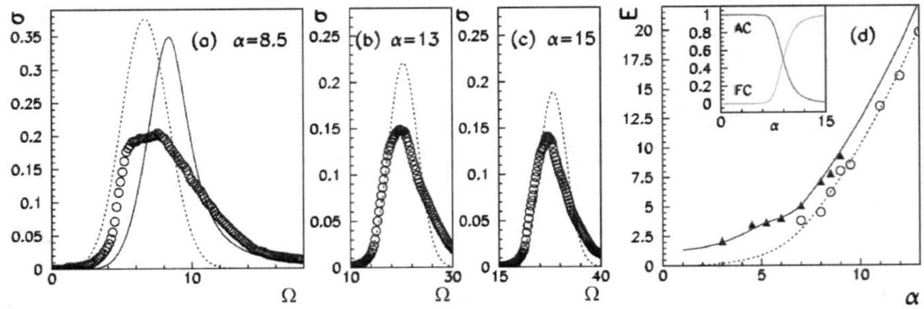

Fig. 2.15. (**a**)–(**c**) Comparison of the optical conductivity calculated with the DQMC method (*circles*), the extended MFF (*solid line*), and SCE (*dashed line*) for three different values of α. (**d**) The energy of the lower- and higher-frequency features obtained by DQMC (*circles and triangles*, respectively) compared (1) with the FC transition energy calculated from the SCE (*dashed line*) and (2) with the energy of the peak obtained from the extended MFF (*solid line*). In the *inset*, the weights of Franck–Condon and adiabatically connected transitions are shown as a function of α (reprinted with permission from [101]. ©2006 by the American Physical Society)

DQMC data [100], showing significant improvement with respect to the weak-coupling perturbation approach [91, 95] which provides a good description of the OC spectra only for very small values of α. For $3 \leq \alpha \leq 6$, DSG predicts the essential structure of the optical absorption, with a RES transition gradually building up for increasing α, but underestimates the peak width. The damping, introduced in the extended MFF approach, becomes crucial in this coupling regime.

Second, comparing the peak and shoulder energies, obtained by DQMC, with the peak energies, given by MFF, and the FC transition energies from the SCE, it is concluded [101] that as α increases from 6 to 10 the spectral weights rapidly switch from the dynamic regime, where the lattice follows the electron motion, to the adiabatic regime dominated by FC transitions. In the intermediate electron–phonon coupling regime, $6 < \alpha < 10$, both adiabatic FC and nonadiabatic dynamical excitations coexist. For still larger coupling $\alpha \gtrsim 10$, the absorption spectrum consists of a broad FC-structure, built of LO-phonon sidebands.

In summary, the accurate numerical results obtained from DQMC and from the recent analytical approximations [67, 101] confirm the essence of the mechanism for the optical absorption of Fröhlich polarons, proposed in [73, 93] combined with [98] and do add important new insights.

2.4.7 Sum Rules for the Optical Conductivity Spectra of Fröhlich Polarons

In this section, several sum rules for the optical conductivity spectra of Fröhlich polarons are applied to test the DSG approach [73] and the DQMC results [100]. The values of the polaron effective mass for the DQMC approach are taken from [55]. In Tables 2.2 and 2.3, we show the polaron ground state E_0 and the zero, M_0 and first, M_1, frequency moments calculated using the optical conductivity spectra:

$$M_0 \equiv \int_1^{\Omega_{\max}} \mathrm{Re}\,\sigma(\Omega)\,\mathrm{d}\Omega, \tag{2.102}$$

$$M_1 \equiv \int_1^{\Omega_{\max}} \Omega \mathrm{Re}\,\sigma(\Omega)\,\mathrm{d}\Omega, \tag{2.103}$$

where $\Omega_{\max}$ is the upper value of the frequency available from [100], and

$$\tilde{M}_0 \equiv \frac{\pi}{2m^*} + \int_1^{\Omega_{\max}} \mathrm{Re}\,\sigma(\Omega)\,\mathrm{d}\Omega. \tag{2.104}$$

Here, m^* is the polaron mass, the optical conductivity is calculated in units $n_\mathrm{p}e^2/(m\omega_0)$, m^* is measured in units of the band mass m, and the frequency

Table 2.2. Polaron parameters $M_0, M_1, \tilde{M}_0$ obtained from the diagrammatic Monte Carlo results (reprinted with permission from [94]. © 2006, Società Italiana di Fisica)

α	$M_0^{(\mathrm{DQMC})}$	$m^{*(\mathrm{DQMC})}$	$\tilde{M}_0^{(\mathrm{DQMC})}$	$M_1^{(\mathrm{DQMC})}/\alpha$	$E_0^{(\mathrm{DQMC})}$
0.01	0.00249	1.0017	1.5706	0.634	−0.010
1	0.24179	1.1865	1.5657	0.65789	−1.013
3	0.67743	1.8467	1.5280	0.73123	−3.18
4.5	0.97540	2.8742	1.5219	0.862	−4.97
5.25	1.0904	3.8148	1.5022	0.90181	−5.68
6	1.1994	5.3708	1.4919	0.98248	−6.79
6.5	1.30	6.4989	1.5417	1.1356	−7.44
7	1.3558	9.7158	1.5175	1.2163	−8.31
8	1.4195	19.991	1.4981	1.3774	−9.85

Table 2.3. Polaron parameters $M_0, M_1, \tilde{M}_0$ obtained within the path-integral approach (reprinted with permission from [94]. © 2006, Società Italiana di Fisica)

α	$M_0^{(\mathrm{DSG})}$	$m^{*(\mathrm{Feynman})}$	$\tilde{M}_0^{(\mathrm{DSG})}$	$M_1^{(\mathrm{DSG})}/\alpha$	$E_0^{(\mathrm{Feynman})}$
0.01	0.00248	1.0017	1.5706	0.633	−0.010
1	0.24318	1.1957	1.5569	0.65468	−1.0130
3	0.69696	1.8912	1.5275	0.71572	−3.1333
4.5	1.0162	3.1202	1.5196	0.83184	−4.8394
5.25	1.1504	4.3969	1.5077	0.88595	−5.7482
6	1.2608	6.8367	1.4906	0.95384	−6.7108
6.5	1.3657	9.7449	1.5269	1.1192	−7.3920
7	1.4278	14.395	1.5369	1.2170	−8.1127
8	1.4741	31.569	1.5239	1.4340	−9.6953

is measured in units of ω_0. The values of $\Omega_{\max}$ are $\Omega_{\max} = 10$ for $\alpha = 0.01$, 1, and 3; $\Omega_{\max} = 12$ for $\alpha = 4.5$, 5.25, and 6; and $\Omega_{\max} = 18$ for $\alpha = 6.5$, 7, and 8.

The optical conductivity derived by DSG [73] exactly satisfies the sum rule [118]:

$$\frac{\pi}{2m^*} + \int_1^\infty \mathrm{Re}\,\sigma(\Omega)\,\mathrm{d}\Omega = \frac{\pi}{2}. \tag{2.105}$$

Since the optical conductivity obtained from the DQMC results [100] is known only within a limited interval of frequencies $1 < \Omega < \Omega_{\max}$, the integral in (2.104) for the DSG approach [73] is calculated over the same frequency interval as for the Monte Carlo results [100].

The comparison of the resulting zero-frequency moments $\tilde{M}_0^{(\mathrm{DQMC})}$ and $\tilde{M}_0^{(\mathrm{DSG})}$ with each other and with the value $\pi/2 = 1.5707963\ldots$ corresponding to the right-hand side of the sum rule (2.105) shows that the difference $\left|\tilde{M}_0^{(\mathrm{DQMC})} - \tilde{M}_0^{(\mathrm{DSG})}\right|$ on the interval $\alpha \leq 8$ is smaller than the absolute value of the contribution of the "tail" of the optical conductivity for $\Omega > \Omega_{\max}$ to the integral in the sum rule (2.105):

$$\int_{\Omega_{\max}}^\infty \mathrm{Re}\,\sigma^{(\mathrm{DSG})}(\Omega)\,\mathrm{d}\Omega \equiv \frac{\pi}{2} - \tilde{M}_0^{(\mathrm{DSG})}. \tag{2.106}$$

Within the accuracy determined by the neglect of the "tail" of the part of the spectrum for $\Omega > \Omega_{\max}$, the contribution to the integral in the sum rule (2.105) for the optical conductivity obtained from the DQMC results [100] *agrees well* with that for the optical conductivity found within the path-integral approach in [73]. Hence, the conclusion follows that *the optical conductivity obtained from the DQMC results* [100] *satisfies the sum rule* (2.105) within the aforementioned accuracy.

We analyze the fulfillment of the "LSD" polaron ground-state theorem introduced in [119]:

$$E_0(\alpha) - E_0(0) = -\frac{3}{\pi}\int_0^\alpha \frac{\mathrm{d}\alpha'}{\alpha'}\int_0^\infty \Omega\mathrm{Re}\,\sigma(\Omega,\alpha')\,\mathrm{d}\Omega \tag{2.107}$$

using the first-frequency moments $M_1^{(\mathrm{DQMC})}$ and $M_1^{(\mathrm{DSG})}$. The results of this comparison are presented in Fig. 2.16. The dots indicate the polaron ground-state energy calculated by Feynman using his variational principle for path integrals. The solid curve is the value of $E_0(\alpha)$ calculated numerically using the optical conductivity spectra and the ground-state theorem with the DSG optical conductivity [73] for the polaron:

$$E_0^{(\mathrm{DSG})}(\alpha) \equiv -\frac{3}{\pi}\int_0^\alpha \frac{\mathrm{d}\alpha'}{\alpha'}\int_0^\infty \Omega\mathrm{Re}\,\sigma^{(\mathrm{DSG})}(\Omega,\alpha')\,\mathrm{d}\Omega. \tag{2.108}$$

The dotted curve and the open circles are the values obtained using $M_1^{(\mathrm{DSG})}(\alpha)$ and $M_1^{(\mathrm{DQMC})}(\alpha)$, respectively:

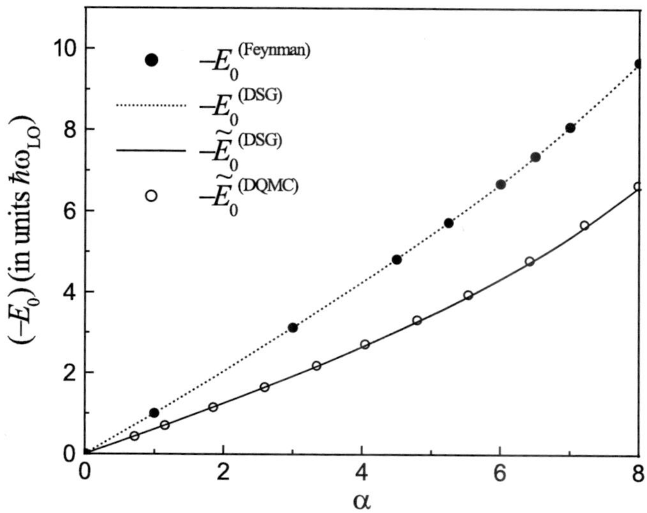

Fig. 2.16. Test of the ground-state theorem for a Fröhlich polaron from [119] using different optical conductivity spectra, DSG from [73], and DQMC from [100]. The notations are explained in the text (reprinted with permission after [94]. © 2006, Società Italiana di Fisica)

$$\tilde{E}_0(\alpha) \equiv -\frac{3}{\pi}\int_0^{\alpha}\frac{\mathrm{d}\alpha'}{\alpha'}\int_0^{\Omega_{\max}}\Omega\operatorname{Re}\sigma(\Omega,\alpha')\,\mathrm{d}\Omega = -\frac{3}{\pi}\int_0^{\alpha}\mathrm{d}\alpha'\frac{M_1(\alpha')}{\alpha'}. \tag{2.109}$$

As seen from the figure, $E_0^{(\text{DSG})}(\alpha)$ coincides, to a high degree of accuracy, with the variational polaron ground-state energy. From the comparison of $\tilde{E}_0^{(\text{DSG})}(\alpha)$ with $\tilde{E}_0^{(\text{DQMC})}(\alpha)$, it follows that the contribution to the integral in (2.109), with the given limited frequency region, which approximates the integral in the right-hand side of the "LSD" ground-state theorem (2.107), for the optical conductivity obtained from the DQMC results [100] *agrees with a high accuracy* with the corresponding contribution to the integral in (2.109) for the optical conductivity derived from the path-integral approach of [73]. Because for the path-integral result, the integral $\tilde{E}_0^{(\text{DSG})}(\alpha)$ noticeably differs from the integral $E_0^{(\text{DSG})}(\alpha)$, a comparison between the Feynman polaron ground-state energy E_0 and the integral $\tilde{E}_0^{(\text{DSG})}(\alpha)$ is not justified. Similarly, a comparison between the polaron ground-state energy obtained from the DQMC results and the integral $\tilde{E}_0^{(\text{DQMC})}(\alpha)$ would require to overcome the limited frequency domain of the available optical conductivity spectrum [100].

The DQMC optical conductivity spectrum for higher frequencies than $\Omega_{\max}$ of [100] is needed to check the fulfillment of the sum rules (2.105) and (2.107) with a higher accuracy.

2.5 Polaron Scaling Relations

The form of the Fröhlich Hamiltonian (2.16) in n-dimensions is the same as in 3D, except that now all vectors are n-dimensional. In this section, we take $m = \omega_0 = 1$. In 3D, the EPI matrix element is well known, $|V_{\mathbf{q}}|^2 = 2\sqrt{2}\pi\alpha/V_3 q^2$. The interaction coefficient in n-dimensions becomes [120]

$$|V_{\mathbf{q}}|^2 = \frac{2^{n-3/2}\pi^{(n-1)/2}\Gamma\left(\frac{n-1}{2}\right)\alpha}{V_n q^{n-1}} \tag{2.110}$$

with V_n the volume of the n-dimensional crystal.

The only difference between the model system in n-dimensions and the model system in 3D is that now one deals with an n-dimensional harmonic oscillator. Directly following [12], the variational polaron energy was calculated in [120]:

$$\begin{aligned} E &= \frac{n(v-w)}{2} - \frac{n\left(v^2-w^2\right)}{4v} - \frac{2^{-3/2}\Gamma\left(\frac{n-1}{2}\right)\alpha}{\Gamma\left(\frac{n}{2}\right)}\int_0^\infty \frac{\mathrm{e}^{-t}}{\sqrt{D_0(t)}}\mathrm{d}t \\ &= \frac{n(v-w)^2}{4v} - \frac{\Gamma\left(\frac{n-1}{2}\right)\alpha}{2\sqrt{2}\Gamma\left(\frac{n}{2}\right)}\int_0^\infty \frac{\mathrm{e}^{-t}}{\sqrt{D_0(t)}}\mathrm{d}t, \end{aligned} \tag{2.111}$$

where

$$D_0(t) = \frac{w^2}{2v^2}t + \frac{v^2-w^2}{2v^3}\left(1-\mathrm{e}^{-vt}\right). \tag{2.112}$$

To facilitate a comparison of E for n-dimensions with the Feynman result [12] for 3D,

$$E_{3\mathrm{D}}(\alpha) = \frac{3(v-w)^2}{4v} - \frac{1}{\sqrt{2\pi}}\alpha\int_0^\infty \frac{\mathrm{e}^{-t}}{\sqrt{D_0(t)}}\mathrm{d}t, \tag{2.113}$$

it is convenient to rewrite (2.111) in the form

$$E_{n\mathrm{D}}(\alpha) = \frac{n}{3}\left[\frac{3(v-w)^2}{4v} - \frac{1}{\sqrt{2\pi}}\frac{3\sqrt{\pi}\Gamma\left(\frac{n-1}{2}\right)}{2n\Gamma\left(\frac{n}{2}\right)}\alpha\int_0^\infty \frac{\mathrm{e}^{-t}}{\sqrt{D_0(t)}}\mathrm{d}t\right]. \tag{2.114}$$

The parameters w and v must be determined by minimizing E. In the case of (2.114), one should minimize the expression in the square brackets. The only difference of this expression from the RHS of (2.113) is that α is multiplied by the factor

$$a_n = \frac{3\sqrt{\pi}\Gamma\left(\frac{n-1}{2}\right)}{2n\Gamma\left(\frac{n}{2}\right)}. \tag{2.115}$$

This means that the minimizing parameters w and v in nD at a given α will be exactly the same as those calculated in 3D with the Fröhlich constant chosen as $a_n\alpha$:

$$v_{n\mathrm{D}}(\alpha) = v_{3\mathrm{D}}(a_n\alpha), \quad w_{n\mathrm{D}}(\alpha) = w_{3\mathrm{D}}(a_n\alpha). \tag{2.116}$$

Comparing (2.114) to (2.113), the following scaling relation [120–122] is obtained:

$$E_{n\mathrm{D}}(\alpha) = \frac{n}{3} E_{3\mathrm{D}}(a_n \alpha), \tag{2.117}$$

where a_n is given by (2.115). As discussed in [120], the above scaling relation is not an *exact* relation. It is valid for the Feynman polaron energy and also for the ground-state energy to order α. The next-order term (i.e., α^2) no longer satisfies (2.117). The reason is that in the exact calculation (to order α^2), the electron motions in different space directions are coupled by EPI. No such coupling appears in the Feynman polaron model; this is the underlying reason for the validity of the scaling relation for the Feynman approximation.

In [82,118,120,121], scaling relations were obtained also for the impedance function, $Z_{n\mathrm{D}}(\alpha;\Omega) = Z_{3\mathrm{D}}(a_n\alpha;\Omega)$, the effective mass, and the mobility of a polaron. In the important particular case of 2D, the scaling relations take the form [120–122]:

$$E_{2\mathrm{D}}(\alpha) = \frac{2}{3} E_{3\mathrm{D}}\left(\frac{3\pi}{4}\alpha\right), \tag{2.118}$$

$$Z_{2\mathrm{D}}(\alpha;\Omega) = Z_{3\mathrm{D}}\left(\frac{3\pi}{4}\alpha;\Omega\right), \tag{2.119}$$

$$\frac{m^*_{2\mathrm{D}}(\alpha)}{(m)_{n\mathrm{D}}} = \frac{m^*_{3\mathrm{D}}\left(\frac{3\pi}{4}\alpha\right)}{(m)_{3\mathrm{D}}}, \tag{2.120}$$

$$\mu_{2\mathrm{D}}(\alpha) = \mu_{3\mathrm{D}}\left(\frac{3\pi}{4}\alpha\right). \tag{2.121}$$

The fulfillment of the scaling relation [121] (PD) can be checked for the path-integral Monte Carlo results [65] for the polaron free energy. The path-integral Monte Carlo results of [65] in 3D and 2D are given for a few values of temperature and for some selected values of α. For a check of the scaling relation, the values of the polaron free energy at $\beta = 10$ ($\beta = \omega_0/T$) are taken from [65] and plotted in Fig. 2.17, upper panel for 2D and 3D, with squares and open circles, respectively. In Fig. 2.17, lower panel, the available data for the free energy from [65] are plotted in the following form, inspired by the LHS and the RHS of (2.118): $F_{2\mathrm{D}}(\alpha)$ (squares) and $(2/3)F_{3\mathrm{D}}(3\pi\alpha/4)$ (open triangles). As follows from the figure, the path-integral Monte Carlo results for the polaron free energy in 2D and 3D *very closely* follow the PD-scaling relation of the form given by (2.118):

$$F_{2\mathrm{D}}(\alpha) \equiv \frac{2}{3} F_{3\mathrm{D}}\left(\frac{3\pi\alpha}{4}\right). \tag{2.122}$$

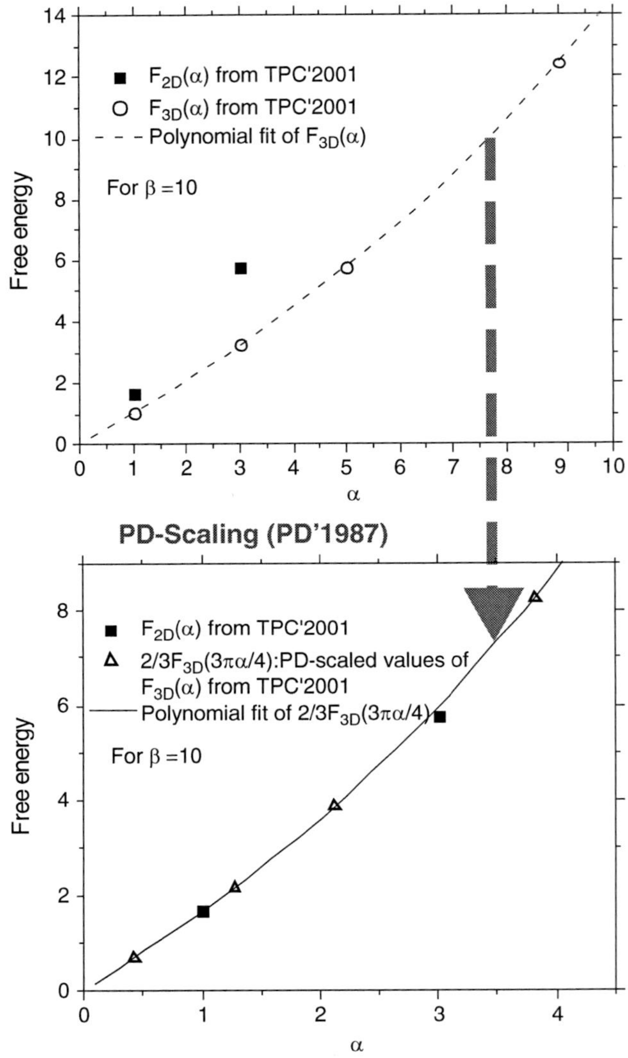

Fig. 2.17. *Upper panel*: the polaron free energy in 2D (*squares*) and 3D (*open circles*) obtained by TPC'2001 [65] for $\beta = 10$. The data for $F_{3D}(\alpha)$ are interpolated using a polynomial fit to the available four points (*dotted line*). *Lower panel*: demonstration of the PD-scaling (cf. PD'1987). The polaron free energy in 2D obtained by TPC'2001 [65] for $\beta = 10$ (*squares*). The PD-scaled according to PD'1987 [121] polaron free energy in 3D from TPC'2001 for $\beta = 10$ (*open triangles*). The data for $(2/3)F_{3D}(3\pi\alpha/4)$ are interpolated using a polynomial fit to the available four points (*solid line*) (reprinted with permission from [94]. © 2006, Società Italiana di Fisica)

3

Lattice Polaron

When the coupling with phonons increases, the polaron radius decreases and becomes of the order of the lattice constant. Then, all momenta of the Brillouin zone contribute to the polaron wave function and the effective mass approximation cannot be applied. This regime occurs if the characteristic potential energy E_{p} (polaron level shift) due to the local lattice deformation is compared or larger than the half-bandwidth D. The strong-coupling regime with the dimensionless coupling constant

$$\lambda \equiv \frac{E_{\mathrm{p}}}{D} \geq 1 \tag{3.1}$$

is called the *small* or *lattice* polaron. In general, E_{p} is expressed as

$$E_{\mathrm{p}} = \frac{1}{2N} \sum_{\mathbf{q}} |\gamma(\mathbf{q})|^2 \omega_{\mathbf{q}} \tag{3.2}$$

for any type of phonons involved in the polaron cloud. For the Fröhlich interaction with optical phonons, one obtains $E_{\mathrm{p}} \simeq q_{\mathrm{d}} e^2/\pi\kappa$, where q_{d} is the Debye momentum [59]. For example, with parameters appropriate for high T_{c} copper oxides $\epsilon_0 \gg \epsilon \simeq 5$ and $q_{\mathrm{D}} \simeq 0.7\,\text{Å}^{-1}$, one obtains $E_{\mathrm{p}} \simeq 0.6\,\mathrm{eV}$ [123, 124]. The exact value of λ_{c} when the continuum (large) polaron transforms into the small one depends on the lattice structure, phonon frequency dispersions, and the radius of the electron–phonon interaction, but in most cases the transformation occurs around $\lambda_{\mathrm{c}} \simeq 1$ [125]. Lattice polarons are expected to be the carriers in oxides, which are strongly polarizable doped semiconductors, if the bare-electron band is narrow enough [26], and in molecular nanowires (Sect. 6.3.2).

3.1 Holstein Model

Main features of small polarons are revealed in the simple Holstein model [18] of two vibrating molecules and the electron hopping between them. A simplified version of the model is defined by a two-site Hamiltonian describing the

A.S. Alexandrov and J.T. Devreese, *Advances in Polaron Physics*, Springer Series in Solid-State Sciences 159,
DOI: 10.1007/978-3-642-01896-1_3, © Springer-Verlag Berlin Heidelberg 2010

electron tunneling between sites 1 ("left") and 2 ("right") with the amplitude t and interacting with a vibrational mode of an ion, placed at some distance in between (Fig. 3.1):

$$H = t(c_1^\dagger c_2 + c_2^\dagger c_1) + H_{\mathrm{ph}} + H_{\mathrm{e\text{-}ph}}. \tag{3.3}$$

Here, we take the position of an atomic level in the rigid lattice as zero, and c_i annihilates the electron on the left, $i = 1$, or on the right, $i = 2$, site.

The vibration part of the Hamiltonian in this toy model is

$$H_{\mathrm{ph}} = -\frac{1}{2M}\frac{\partial^2}{\partial x^2} + \frac{kx^2}{2}, \tag{3.4}$$

where M is the ion mass, $k = M\omega_0^2$ is the spring constant, and x is the ion displacement. The electron–phonon interaction, $H_{\mathrm{e\text{-}ph}}$, depends on the polarization of vibrations. If the ion vibrates along the perpendicular direction to the hopping (in c-direction, Fig. 3.1), we have

$$H_{\mathrm{e\text{-}ph}} = f_c x(c_1^\dagger c_1 + c_2^\dagger c_2) \tag{3.5}$$

and

$$H_{\mathrm{e\text{-}ph}} = f_a x(c_1^\dagger c_1 - c_2^\dagger c_2), \tag{3.6}$$

if the ion vibrates along the hopping (a-direction).

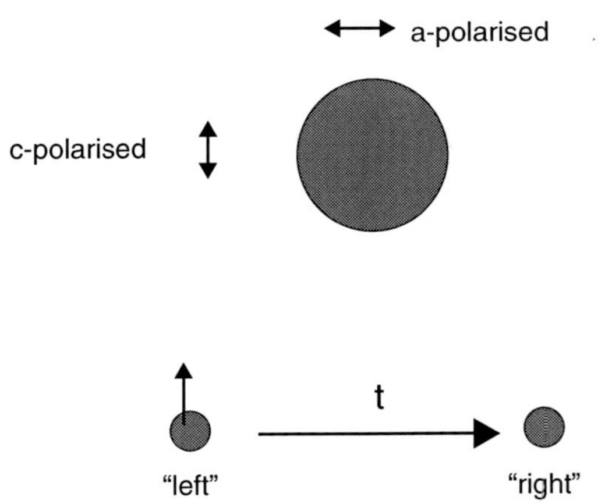

Fig. 3.1. Electron tunnels between sites 1 ("left") and 2 ("right") with the amplitude t and interacts with c-axis or a-axis polarized vibrational modes of the ion, placed in between

The wave function of the electron and the ion is a linear superposition of two terms describing the electron on the "left" and "right" site, respectively:

$$\psi = [u(x)c_1^\dagger + v(x)c_2^\dagger]|0\rangle, \tag{3.7}$$

where $|0\rangle$ is the vacuum state describing a rigid lattice without the extra electron. Substituting ψ into the Schrödinger equation, $H\psi = E\psi$, we obtain two coupled equations for the amplitudes:

$$(E - f_{a,c}x - H_{\text{ph}})u(x) = tv(x), \tag{3.8}$$

$$(E \pm f_{a,c}x - H_{\text{ph}})v(x) = tu(x), \tag{3.9}$$

where + and − in the second equation correspond to a- and c-polarized vibrations, respectively. There is the exact solution for the c-axis polarization, when a change in the ion position leads to the same shift of the electron energy on the left and right sites:

$$u(x) = u\chi_n(x),$$
$$v(x) = v\chi_n(x), \tag{3.10}$$

where u and v are constants and

$$\chi_n(x) = \left(\frac{M\omega_0}{\pi(2^n n!)^2}\right)^{1/4} H_n[(x - f_c/k)(M\omega_0)^{1/2}] \exp[-M(x - f_c/k)^2\omega_0/2] \tag{3.11}$$

is the harmonic oscillator wave function. There are two ladders of levels given by

$$E_n^{\pm} = -E_{\text{p}} \pm t + \omega_0(n + 1/2) \tag{3.12}$$

with $E_{\text{p}} = f_c^2/2k$. Here, $H_n(\xi)$ are the Hermite polynomials, and $n = 0, 1, 2, 3, \ldots$. Hence, the c-axis single-ion deformation leads to the polaron level shift but without any renormalization of the hopping integral t. In contrast, a-polarized vibrations with the opposite shift of the electron energy on the left and right sites strongly renormalize the hopping integral. There is no simple general solution of the Holstein model in this case, but one can find it in two limiting cases, *nonadiabatic*, when $t \ll \omega_0$ and *adiabatic*, when $t \gg \omega_0$.

3.1.1 Nonadiabatic Holstein Polaron

In the nonadiabatic regime, the ion vibrations are fast and the electron hopping is slow. Hence, one can apply a perturbation theory in powers of t to solve

$$\begin{pmatrix} E - f_a x - H_{\text{ph}} & -t \\ -t & E + f_a x - H_{\text{ph}} \end{pmatrix} \begin{bmatrix} u(x) \\ v(x) \end{bmatrix} = 0. \tag{3.13}$$

One takes $t = 0$ in zero order and obtains a twofold *degenerate* ground state $[u^{\mathrm{l,r}}(x), v^{\mathrm{l,r}}(x)]$, corresponding to the polaron localized on the left (l) or on the right (r) sites:

$$u^{\mathrm{l}}(x) = \exp\left[-\frac{M\omega_0}{2}(x + f_a/k)^2\right], \tag{3.14}$$

$$v^{\mathrm{l}}(x) = 0 \tag{3.15}$$

and

$$u^{\mathrm{r}}(x) = 0, \tag{3.16}$$

$$v^{\mathrm{r}}(x) = \exp\left[-\frac{M\omega_0}{2}(x - f_a/k)^2\right] \tag{3.17}$$

with the energy $E_0 = -E_{\mathrm{p}} + \omega_0/2$, where $E_{\mathrm{p}} = f_a^2/2k$. The eigenstates are found as linear superpositions of two unperturbed states:

$$\begin{bmatrix} u(x) \\ v(x) \end{bmatrix} = \alpha \begin{bmatrix} u^{\mathrm{l}}(x) \\ 0 \end{bmatrix} + \beta \begin{bmatrix} 0 \\ v^{\mathrm{r}}(x) \end{bmatrix}. \tag{3.18}$$

Here, the coefficients α and β are independent of x. The conventional secular equation for E is obtained, multiplying the first row by $u^{\mathrm{l}}(x)$ and the second row by $v^{\mathrm{r}}(x)$, and integrating over the vibration coordinate, x, each of two equations of the system. The result is

$$\det\begin{pmatrix} E - E_0 & \tilde{t} \\ \tilde{t} & E - E_0 \end{pmatrix} = 0 \tag{3.19}$$

with the renormalized hopping integral

$$\frac{\tilde{t}}{t} = \frac{\int_{-\infty}^{\infty} \mathrm{d}x u^{\mathrm{l}}(x) v^{\mathrm{r}}(x)}{\int_{-\infty}^{\infty} \mathrm{d}x |u^{\mathrm{l}}(x)|^2}. \tag{3.20}$$

The corresponding eigenvalues $E_{\pm}$ are

$$E_{\pm} = \omega_0/2 - E_{\mathrm{p}} \pm \tilde{t}. \tag{3.21}$$

The hopping integral splits the degenerate level, as in the rigid lattice, but an effective "bandwidth" $2\tilde{t}$ is significantly reduced compared with the bare one:

$$\tilde{t} = t \exp(-2E_{\mathrm{p}}/\omega_0). \tag{3.22}$$

This polaron band narrowing originates in a small overlap integral of two displaced oscillator wave functions $u^{\mathrm{l}}(x)$ and $v^{\mathrm{r}}(x)$.

3.1.2 Adiabatic Holstein Polaron

In the adiabatic regime, when $t \gg \omega_0$, the electron tunneling is fast compared with the ion motion. Hence, one can apply the Born–Oppenheimer adiabatic approximation [2] taking the wave function in the form

$$\begin{bmatrix} u(x) \\ v(x) \end{bmatrix} = \chi(x) \begin{bmatrix} u_a(x) \\ v_a(x) \end{bmatrix}. \tag{3.23}$$

Here, $u_a(x)$ and $v_a(x)$ are the electron wave functions obeying the Schrödinger equation with the *frozen* ion deformation x, i.e.,

$$\begin{pmatrix} E_a(x) - f_a x & -t \\ -t & E_a(x) + f_a x \end{pmatrix} \begin{bmatrix} u_a(x) \\ v_a(x) \end{bmatrix} = 0. \tag{3.24}$$

The lowest energy level is found as

$$E_a(x) = -\sqrt{(f_a x)^2 + t^2}. \tag{3.25}$$

$E_a(x)$ together with $kx^2/2$ plays the role of a potential energy term in the equation for the "vibration" wave function, $\chi(x)$:

$$\left[-\frac{1}{2M}\frac{\partial^2}{\partial x^2} + \frac{kx^2}{2} - \sqrt{(f_a x)^2 + t^2} \right] \chi(x) = E\chi(x). \tag{3.26}$$

Terms with the first and second derivatives of the electron wave functions $u_a(x)$ and $v_a(x)$ are small compared with the corresponding derivatives of $\chi(x)$ in the adiabatic approximation, so they are neglected in (3.26). As a result, we arrive at the familiar double-well potential problem, where the potential energy $U(x) = kx^2/2 - \sqrt{(f_a x)^2 + t^2}$ has two symmetric minima, separated by a barrier. Minima are located approximately at

$$x_{\mathrm{m}} = \pm f_a/k \tag{3.27}$$

in the strong-coupling limit, $E_{\mathrm{p}} \gg t$, and the potential energy near the bottom of each potential well is about

$$U(x) = -E_{\mathrm{p}} + \frac{k(|x| - f_a/k)^2}{2}. \tag{3.28}$$

If the barrier were impenetrable, there would be the ground-state energy level $E_0 = -E_{\mathrm{p}} + \omega_0/2$, the same for both wells. The underbarrier tunneling results in a splitting of this level $2\tilde{t}$, which corresponds to a polaron bandwidth in the lattice. It can be estimated using the quasiclassical approximation as

$$\tilde{t} \propto \exp\left[-2\int_0^{x_{\mathrm{m}}} p(x)\mathrm{d}x \right], \tag{3.29}$$

where $p(x) = \sqrt{2M[U(x) - E_0]} \approx (Mk)^{1/2}|x - f_a/k|$ is the classical momentum.

Estimating the integral, one finds the exponential reduction of the "bandwidth":

$$\tilde{t} \propto \exp(-2E_{\mathrm{p}}/\omega_0), \tag{3.30}$$

which is the same as in the nonadiabatic regime. Holstein found corrections to this expression up to terms of the order of $1/\lambda^2$, which allowed him to improve the exponent and estimate the pre-exponential factor as

$$\tilde{t} \approx \sqrt{\frac{E_{\mathrm{p}}\omega_0}{\pi}} \mathrm{e}^{-\tilde{g}^2}. \tag{3.31}$$

Here, $g^2 = 2E_{\mathrm{p}}/\omega_0$ and

$$\tilde{g}^2 = g^2 \left[1 - \frac{1}{4\lambda^2} \ln(4\lambda) - \frac{1}{8\lambda^2}\right]. \tag{3.32}$$

A more accurate expression for $\tilde{t}$ was obtained in [126]:

$$\tilde{t} \approx \sqrt{\frac{E_{\mathrm{p}}\omega_0}{\pi}} \beta^{5/2} \lambda^{1-\beta} [2(1+\beta)]^{-\beta} \mathrm{e}^{-\tilde{g}^2}, \tag{3.33}$$

where now $\tilde{g}^2 = g^2\{\beta - [\ln(2\lambda(1+\beta))]/4\lambda^2\}$ and $\lambda = E_{\mathrm{p}}/t$. This expression takes into account the phonon frequency renormalization, $\beta \equiv \tilde{\omega}_0/\omega_0 = \sqrt{1 - 1/4\lambda^2}$, and the anharmonic corrections of the order of $1/\lambda^2$ to the turning point x_{m} in (3.29). The term in front of the exponent in (3.31) and (3.33) differs from t of the nonadiabatic case (3.22). It is thus apparent that the perturbation approach covers only a part of the entire lattice polaron region, $\lambda \gtrsim 1$. The upper limit of applicability of the perturbation theory is given by $t < \sqrt{E_{\mathrm{p}}\omega_0}$. For the remainder of the region, the adiabatic approximation is more appropriate. A much lower effective mass of the adiabatic small polaron in the intermediate-coupling region compared with that estimated from the perturbation theory expression (3.22) is revealed in (3.33) [126]. The double-well potential disappears at $\lambda = \lambda_{\mathrm{c}} = 0.5$, where the renormalized phonon frequency $\tilde{\omega}_0$ is zero.

The Holstein polaron model can be readily generalized for infinite lattices (Sect. 3.2). Similar models were used, for instance, in studies of dissipation [127, 128] and effects of decoherence in open quantum-mechanical systems.

3.2 Lang–Firsov Canonical Transformation

The kinetic energy is smaller than the interaction energy as long as $\lambda > 1$. Hence, a self-consistent approach to the lattice polaron problem is possible with the "$1/\lambda$" expansion technique [19, 129] on infinite lattices with any type of EPI conserving the electron site occupation numbers, and any phonon spectrum. The technique treats the electron kinetic energy as a perturbation, and can be applied for multipolaron systems as well [130] (see Sect. 4.2). It is based on the fact, known for a long time, that there is an analytical exact solution of the *any-number* polaron problem in the extreme strong-coupling limit, $\lambda \to \infty$. Following Lang and Firsov, one applies the canonical transformation e^S to diagonalize the Hamiltonian (1.32). The diagonalization is *exact*, if $t(\mathbf{m}) = 0$ (or $\lambda = \infty$):

$$\tilde{H} = \mathrm{e}^S H \mathrm{e}^{-S}, \tag{3.34}$$

where

$$S = -\sum_{\mathbf{q},i} \hat{n}_i \left[u_i(\mathbf{q})d_{\mathbf{q}} - \text{H.c.}\right] \tag{3.35}$$

is such that $S^\dagger = -S$. The electron and phonon operators are transformed as $\tilde{c}_i = \mathrm{e}^S c_i \mathrm{e}^{-S}$ and $\tilde{d}_{\mathbf{q}} = \mathrm{e}^S d_{\mathbf{q}} \mathrm{e}^{-S}$. The result is

$$\tilde{c}_i = c_i \hat{X}_i \tag{3.36}$$

and

$$\tilde{d}_{\mathbf{q}} = d_{\mathbf{q}} - \sum_i \hat{n}_i u_i^*(\mathbf{q}), \tag{3.37}$$

where $\hat{X}_i = \exp\left[\sum_{\mathbf{q}} u_i(\mathbf{q})d_{\mathbf{q}} - \text{H.c.}\right]$. The Lang–Firsov transformation shifts the ions to new equilibrium positions. In a more general sense, it changes the boson vacuum. As a result, the transformed Hamiltonian takes the following form:

$$\tilde{H} = \sum_{i,j} \hat{\sigma}_{ij} c_i^\dagger c_j - E_{\mathrm{p}} \sum_i \hat{n}_i + \sum_{\mathbf{q}} \omega_{\mathbf{q}}(d_{\mathbf{q}}^\dagger d_{\mathbf{q}} + 1/2) + \frac{1}{2}\sum_{i\neq j} v_{ij}\hat{n}_i\hat{n}_j, \tag{3.38}$$

where

$$\hat{\sigma}_{ij} = t(\mathbf{m}-\mathbf{n})\delta_{ss'}\hat{X}_i^\dagger \hat{X}_j \tag{3.39}$$

is the renormalized hopping integral depending on the phonon operators, and

$$v_{ij} = V_{\mathrm{c}}(\mathbf{m}-\mathbf{n}) - \frac{1}{N}\sum_{\mathbf{q}} |\gamma(\mathbf{q})|^2 \omega_{\mathbf{q}} \cos[\mathbf{q}\cdot(\mathbf{m}-\mathbf{n})] \tag{3.40}$$

is the interaction of polarons, $v_{ij} \equiv v(\mathbf{m}-\mathbf{n})$, comprising their Coulomb repulsion and the interaction via the lattice deformation. In the extreme infinite-coupling limit, $\lambda \to \infty$, we can neglect the hopping term of the transformed Hamiltonian. The rest has analytically determined eigenstates and eigenvalues. The eigenstates $|\tilde{N}\rangle = |n_i, n_{\mathbf{q}}\rangle$ are sorted by the polaron $n_{\mathbf{m}s}$ and phonon $n_{\mathbf{q}}$ occupation numbers. The energy levels are

$$E = -(\mu + E_{\mathrm{p}})\sum_i n_i + \frac{1}{2}\sum_{i\neq j} v_{ij} n_i n_j + \sum_{\mathbf{q}} \omega_{\mathbf{q}}(n_{\mathbf{q}} + 1/2), \tag{3.41}$$

where $n_i = 0, 1$ and $n_{\mathbf{q}} = 0, 1, 2, 3, \ldots, \infty$.

3.2.1 "$1/\lambda$" Expansion and Polaron Band

The Hamiltonian $\tilde{H}$ in zero order with respect to the hopping describes localized polarons and independent phonons, which are vibrations of ions relative to new equilibrium positions depending on the polaron occupation numbers. The middle of the electron band is shifted down by the polaron level shift E_p due to the potential well created by lattice deformation. Importantly, the phonon frequencies remain unchanged in this limit at any polaron density, n. At finite λ and n, there is a softening of phonons $\delta\omega_0$ of the order of $\omega_0 n/\lambda^2$ [131–134] (the initial paper on the phonon renormalization [132] predicting $\delta\omega_0 \propto 1/\lambda$ was subsequently corrected [131, 133, 134]). Interestingly, the optical phonon can be mixed with a low-frequency polaronic plasmon forming a new excitation, "plasphon," which was proposed in [131] as an explanation of the anomalous phonon-mode splitting observed in cuprates [135].

Now, let us discuss the $1/\lambda$ expansion. First, we restrict the discussion to a single-polaron problem with no polaron–polaron interaction. The finite hopping term leads to the polaron tunneling because of degeneracy of the zero-order Hamiltonian with respect to the site position of the polaron. To see how the tunneling occurs, we apply the perturbation theory using $1/\lambda$ as a small parameter. The proper Bloch set of N-fold degenerate zero-order eigenstates with the lowest energy $(-E_\mathrm{p})$ of the unperturbed Hamiltonian is

$$|\mathbf{k}, 0\rangle = \frac{1}{\sqrt{N}} \sum_{\mathbf{m}} c^\dagger_{\mathbf{m}s} \exp(\mathrm{i}\mathbf{k}\cdot\mathbf{m})|0\rangle, \tag{3.42}$$

where $|0\rangle$ is the vacuum and N is the number of sites. By applying the textbook perturbation theory, one readily calculates the perturbed energy levels. Up to the second order in the hopping integral, they are given by

$$E_\mathbf{k} = -E_\mathrm{p} + \epsilon_\mathbf{k} - \sum_{\mathbf{k}', n_\mathbf{q}} \frac{|\langle \mathbf{k}, 0| \sum_{i,j} \hat{\sigma}_{ij} c^\dagger_i c_j |\mathbf{k}', n_\mathbf{q}\rangle|^2}{\sum_\mathbf{q} \omega_\mathbf{q} n_\mathbf{q}}, \tag{3.43}$$

where $|\mathbf{k}', n_\mathbf{q}\rangle$ are the exited states of the unperturbed Hamiltonian with one electron and at least one real phonon. The second term in this equation, which is linear with respect to the bare hopping $t(\mathbf{m})$, describes the polaron band dispersion [136]:

$$\epsilon_\mathbf{k} = \sum_\mathbf{m} t(\mathbf{m})\, \mathrm{e}^{-g^2(\mathbf{m})} \exp(-\mathrm{i}\mathbf{k}\cdot\mathbf{m}), \tag{3.44}$$

where

$$g^2(\mathbf{m}) = \frac{1}{2N} \sum_\mathbf{q} |\gamma(\mathbf{q})|^2 [1 - \cos(\mathbf{q}\cdot\mathbf{m})] \tag{3.45}$$

is the *band-narrowing factor* at zero temperature.

The third term, quadratic in $t(\mathbf{m})$, yields a negative $\mathbf{k}$-*independent* correction to the polaron level shift of the order of $1/\lambda^2$, and a small correction to the polaron band dispersion (3.44) [131, 137, 138]. The correction to the level shift due to polaronic hops onto a neighboring site with no deformation around it. As any second-order correction, this transition shifts the energy down by an amount of about $-t^2(\mathbf{m})/E_\mathrm{p}$. It has little to do with the polaron effective mass and the polaron tunneling mobility because the lattice deformation does not follow the electron. The polaron hops back and forth many times (about e^{g^2}) "waiting" for a sufficient lattice deformation to appear around neighboring site $\mathbf{n}$. Only after the deformation around the neighboring site is created does the polaron tunnel onto the next site together with the deformation.

3.2.2 Temperature Effect on the Polaron Band

Let us now analyze the temperature dependence of the polaron bandwidth, which is determined by the average of the multiphonon operator (3.39):

$$\langle\langle \hat{X}_i^\dagger \hat{X}_j \rangle\rangle \equiv \prod_{\mathbf{q}} \langle\langle \exp[u_i^*(\mathbf{q})d_{\mathbf{q}}^\dagger - \text{H.c.}] \exp[u_j(\mathbf{q})d_{\mathbf{q}\nu} - \text{H.c.}] \rangle\rangle. \tag{3.46}$$

Here, the double angular brackets correspond to quantum as well as statistical averages of any operator $\hat{A}$ with the Gibbs distribution:

$$\langle\langle \hat{A} \rangle\rangle = \sum_\nu e^{(\Omega - E_\nu)/T} \langle \nu | \hat{A} | \nu \rangle \equiv \text{Tr}\{e^{(\Omega - \tilde{H})/T} \hat{A}\}, \tag{3.47}$$

where Ω is the thermodynamic potential and $|\nu\rangle$ are the eigenstates of $\tilde{H}$ with the eigenvalues E_ν. An operator identity $\exp(\hat{A} + \hat{B}) = \exp(\hat{A}\exp(\hat{B}))\exp(-[\hat{A}, \hat{B}]/2)$ is instrumental. It is applied when the commutator $[\hat{A}, \hat{B}]$ is a number. The identity allows us to write

$$\begin{aligned} e^{[u_i^*(\mathbf{q})d_{\mathbf{q}}^\dagger - \text{H.c.}]} e^{[u_j(\mathbf{q})d_{\mathbf{q}} - \text{H.c.}]} &= e^{\alpha^* d_{\mathbf{q}}^\dagger} e^{-\alpha d_{\mathbf{q}}} e^{-|\alpha|^2/2} \\ &\times e^{[u_i(\mathbf{q})u_j^*(\mathbf{q}) - u_i^*(\mathbf{q})u_j(\mathbf{q})]/2}. \end{aligned} \tag{3.48}$$

Quantum and statistical averages are calculated by expanding the exponents in the trace as

$$\langle\langle e^{\alpha^* d^\dagger} e^{-\alpha d} \rangle\rangle = (1-p) \sum_{N=0}^{\infty} \sum_{n=0}^{N} p^N (-1)^n \frac{|\alpha|^{2n}}{(n!)^2} N(N-1) \times \cdots \times (N-n+1), \tag{3.49}$$

where we dropped the phonon and site quantum numbers for transparency. Here, $p = \exp(-\omega_{\mathbf{q}\nu}/T)$, so that a single-mode phonon partition function is $Z_{\text{ph}} = 1/(1-p)$. Equation (3.49) can be written in the form [19]

$$\langle\langle e^{\alpha^* d^\dagger} e^{-\alpha d} \rangle\rangle = (1-p) \sum_{n=0}^{N} (-1)^n \frac{|\alpha|^{2n}}{(n!)^2} p^n \frac{d^n}{dp^n} \sum_{N=0}^{\infty} p^N. \tag{3.50}$$

Taking the sum over N, $\sum_{N=0}^{\infty} p^N = 1/(1-p)$, and differentiating it n times yield $n!$ in the numerator, after which the series over n turns out equal to

$$\langle\langle e^{\alpha^* d^\dagger} e^{-\alpha d} \rangle\rangle = e^{-|\alpha|^2 n_\omega}, \tag{3.51}$$

where $n_\omega = [\exp(\omega_{\mathbf{q}}/T) - 1]^{-1}$ is the Bose–Einstein phonon distribution function. Collecting all multipliers, one finally obtains

$$\langle\langle\hat{\sigma}_{ij}\rangle\rangle = T(\mathbf{m}-\mathbf{n})\delta_{ss'} \exp\left(-\frac{1}{2N}\sum_{\mathbf{q}}|\gamma(\mathbf{q})|^2[1-\cos(\mathbf{q}\cdot\mathbf{m})]\coth\frac{\omega_{\mathbf{q}}}{2T}\right), \tag{3.52}$$

with the zero-temperature limit given by (3.44).

The small-polaron band is exponentially narrow. Hence, one can raise a concern about its existence in real solids [139]. At zero temperature, the perturbation term of the transformed Hamiltonian conserves the momentum because all off-diagonal matrix elements vanish:

$$\left\langle \mathbf{k},0 \left| \sum_{i,j} \hat{\sigma}_{i,j} c_i^\dagger c_j \right| \mathbf{k}',0 \right\rangle = 0 \tag{3.53}$$

if $\mathbf{k} \neq \mathbf{k}'$. The emission of a single high-frequency phonon is impossible for any $\mathbf{k}$ because of the energy conservation. The *polaron* half-bandwidth is exponentially reduced,

$$w \approx D\,\mathrm{e}^{-g^2}, \tag{3.54}$$

and it is usually less than the optical phonon energy ω_0 (g^2 is about $D\lambda/\omega_0$). Hence, there is no damping of the polaron band at $T = 0$ caused by optical phonons, no matter how strong the interaction is. The phonons "dress" the electron and coherently follow its motion. However, at finite temperatures the simultaneous emission and absorption of phonons are possible. Moreover, the polaron bandwidth shrinks with increasing temperature because the phonon-averaged hopping integrals depend on temperature (3.52). For high temperatures, $T \gg \omega_0/2$, the band narrows exponentially as $w \approx D\,\mathrm{e}^{-T/T_0}$, where

$$T_0^{-1} = \frac{1}{N}\sum_{\mathbf{q}}|\gamma(\mathbf{q})|^2\omega_{\mathbf{q}}^{-1}[1-\cos(\mathbf{q}\cdot\mathbf{m})].$$

On the other hand, the two-phonon scattering of polarons becomes more important with increasing temperature. One can estimate this scattering rate by applying the Fermi–Dirac golden rule:

$$\frac{1}{\tau} = 2\pi\left\langle \sum_{\mathbf{q},\mathbf{q}'} |M_{\mathbf{q}\mathbf{q}'}|^2\,\delta(\epsilon_{\mathbf{k}} - \epsilon_{\mathbf{k}+\mathbf{q}-\mathbf{q}'})\right\rangle, \tag{3.55}$$

where the corresponding matrix element is

$$M_{\mathbf{q}\mathbf{q}'} = \sum_{i,j}\langle \mathbf{k}+\mathbf{q}-\mathbf{q}', n_{\mathbf{q}}-1, n_{\mathbf{q}'}+1|\,\hat{\sigma}_{i,j}c_i^\dagger c_j\,|\mathbf{k}, n_{\mathbf{q}}, n_{\mathbf{q}'}\rangle .$$

For simplicity, we consider the momentum independent $\gamma(\mathbf{q}) = \gamma_0$ and $\omega_{\mathbf{q}} = \omega_0$. Expanding $\hat{\sigma}_{ij}$-operators in powers of the phonon creation and annihilation operators, one estimates the matrix element of the two-phonon scattering as $M_{\mathbf{qq}'} \approx N^{-1} w \gamma_0^2 \sqrt{n_{\mathbf{q}}(n_{\mathbf{q}'}+1)}$. Using this estimate and the polaron density of states (DOS), $\rho_{\mathrm{p}}(\xi) \equiv N^{-1} \sum_{\mathbf{k}} \delta(\xi - \epsilon_{\mathbf{k}}) \approx 1/2w$, one obtains [26]

$$\frac{1}{\tau} \approx w \gamma_0^4 n_\omega (1 + n_\omega), \tag{3.56}$$

where $n_\omega = [\exp(\omega_0/T) - 1]^{-1}$ is the phonon distribution function.

The polaron band is well defined, if $1/\tau < w$, which is satisfied for a temperature range $T \leq T_{\min} \approx \omega_0/\ln \gamma_0^4$ about half of the characteristic phonon frequency for relevant values of γ_0^2. At higher temperatures, the incoherent thermal activated hopping dominates in the polaron dynamics [18–20,22], and the polaron states are no longer the Bloch states. When the optical phonon frequencies are exceptionally high (i.e., about 1,000 K as in high-temperature superconductors [135]), lattice polarons are in the Bloch states in the relevant range of temperatures, where the Boltzmann kinetic theory with renormalized energy spectrum is applied.

3.3 Effect of EPI Range and Phonon Dispersion on Lattice Polaron Dynamics

The narrowing of the band and the polaron effective mass strongly depend on the range of EPI [59]. Let us compare the small Holstein polaron (SHP) formed by the zero-range EPI and a small polaron formed by the long-range (Fröhlich) interaction, which we refer to as the small Fröhlich polaron (SFP). We use the real-space representation of $H_{\text{e-ph}}$ [60]:

$$H_{\text{e-ph}} = \sum_{\mathbf{n},i} f(\mathbf{m} - \mathbf{n}) \xi_{\mathbf{n}} \hat{n}_i, \tag{3.57}$$

with the normal coordinate at site $\mathbf{n}$

$$\xi_{\mathbf{n}} = \sum_{\mathbf{q}} (2NM\omega_{\mathbf{q}})^{-1/2} \mathrm{e}^{\mathrm{i}\mathbf{q}\cdot\mathbf{n}} d_{\mathbf{q}} + \text{H.c.} \tag{3.58}$$

and the force between the electron at site $\mathbf{m}$ and the normal coordinate $\xi_{\mathbf{n}}$

$$f(\mathbf{m}) = N^{-1} \sum_{\mathbf{q}} \gamma(\mathbf{q}) (M\omega_{\mathbf{q}}^3)^{1/2} \mathrm{e}^{\mathrm{i}\mathbf{q}\cdot\mathbf{m}}. \tag{3.59}$$

In general, there is no simple relation between the polaron level shift E_{p} and the exponent g^2 of the mass enhancement. This relation depends on the form of EPI. Indeed for EPI with a single dispersionless phonon mode, $\omega_{\mathbf{q}} = \omega_0$, one obtains

$$E_{\mathrm{p}} = \frac{1}{2M\omega_0^2} \sum_{\mathbf{m}} f^2(\mathbf{m}) \tag{3.60}$$

and

$$g^2 = \frac{1}{2M\omega_0^3} \sum_{\mathbf{m}} \left[f^2(\mathbf{m}) - f(\mathbf{m}) f(\mathbf{m}+\mathbf{a}) \right] , \qquad (3.61)$$

where $\mathbf{a}$ is the primitive lattice vector. In the nearest-neighbor approximation, the effective mass renormalization is given by $m^*/m = e^{g^2}$, where $1/m^* = \partial^2 \epsilon_{\mathbf{k}}/\partial k^2$ at $k \to 0$ is the inverse polaron mass. If the interaction is short-ranged, $f(\mathbf{m}) = \kappa \delta_{\mathbf{m},0}$ (the Holstein model), then $g^2 = E_\mathrm{p}/\omega_0$. Here, κ is a constant. In general, one has $g^2 = \gamma E_\mathrm{p}/\omega_0$ with the numerical coefficient

$$\gamma = \frac{1 - \sum_{\mathbf{m}} f(\mathbf{m}) f(\mathbf{m}+\mathbf{a})}{\sum_{\mathbf{n}} f^2(\mathbf{n})}, \qquad (3.62)$$

which might be less than 1. To estimate γ, let us consider a one-dimensional chain model with the long-range Coulomb interaction between the electron on one chain ($\mathbf{m}$) and ion vibrations of another chain ($\mathbf{n}$), polarized in the direction perpendicular to the chains [60] (Fig. 3.2). The corresponding force is given by

$$f(\mathbf{m}-\mathbf{n}) = \frac{\kappa}{(|\mathbf{m}-\mathbf{n}|^2 + 1)^{3/2}}. \qquad (3.63)$$

Here, the distance along the chains $|\mathbf{m}-\mathbf{n}|$ is measured in units of the lattice constant a, the interchain distance is also a, and we take $a = 1$. For this long-range interaction, we obtain $E_\mathrm{p} = 1.27\kappa^2/(2M\omega_0^2)$, $g^2 = 0.49\kappa^2/(2M\omega_0^3)$, and $g^2 = 0.39 E_\mathrm{p}/\omega_0$. Thus, the effective mass renormalization in the nonadiabatic regime is much smaller than that in the Holstein model, roughly as $m^*_{\mathrm{SFP}} \propto (m^*_{\mathrm{SHP}})^{1/2}$ in units of m. An analytical solution of a two-site single-electron system interacting with many vibrating ions of a lattice via a long-range Fröhlich EPI found that SFP is also several orders of magnitude lighter than SHP in the adiabatic regime [140].

Another interesting point is that the size of SFP and the length, over which the distortion spreads, are *different*. In the strong-coupling limit, the polaron is almost localized on one site $\mathbf{m}$. Hence, the size of its wave function is the atomic size. On the other hand, the ion displacements, proportional to the displacement force $f(\mathbf{m}-\mathbf{n})$, spread over a large distance. Their amplitude at a site $\mathbf{n}$ falls with the distance as $|\mathbf{m}-\mathbf{n}|^{-3}$ in the one-dimensional model (Fig. 3.2). The polaron cloud (i.e., lattice distortion) is more extended than the polaron itself. Such polaron tunnels with a larger probability than the Holstein polaron due to a smaller *relative* lattice distortion around two neighboring sites. For a short-range EPI, the *entire* lattice deformation disappears at one site and then forms at its neighbor, when the polaron tunnels from site to site. Therefore, $\gamma = 1$ and the polaron is very heavy already at $\lambda \approx 1$. On the contrary, if the interaction is long-ranged, only a fraction of the total deformation changes every time the polaron tunnels from one site to its neighbor, and γ is smaller than 1.

A lighter mass of SFP compared with the nondispersive SHP is a generic feature of any dispersive electron–phonon interaction. For example, a short-range interaction with dispersive acoustic phonons ($\gamma(\mathbf{q}) \propto 1/q^{1/2}$, $\omega_\mathbf{q} \propto q$)

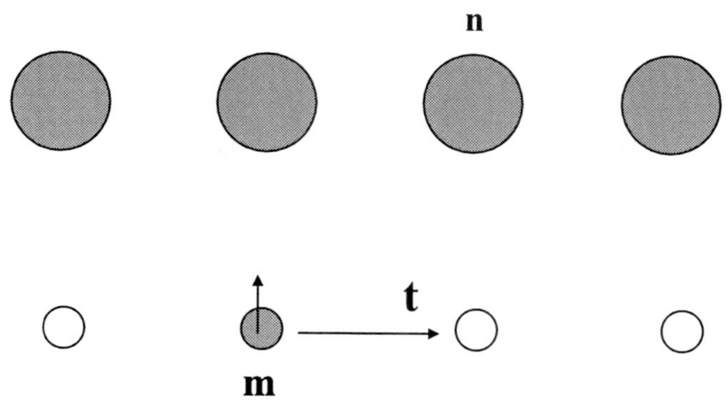

Fig. 3.2. A one-dimensional model of the lattice polaron on chain **m** interacting with displacements of all ions of another chain **n** (t is the polaron hopping integral along chain **m**)

also leads to a lighter polaron in the strong-coupling regime compared with SHP [9]. Even within the Holstein model with the local (intramolecular) EPI, the dispersion of phonon frequencies is a vital ingredient since nondispersive phonons might lead to a divergent site jump probability of polarons [22]. Importantly, the comprehensive studies of the molecular Holstein Hamiltonian, in which the dispersive features of the phonon spectrum are taken into account, found much lower values of the polaron mass compared with the nondispersive model [141–143]. In those studies, the $1/\lambda$ perturbation theory based on the standard Lang–Firsov (LF) and the variational (modified) MLF transformation of the molecular Holstein Hamiltonian with dispersive phonons has been applied for 1D, 2D, and 3D lattices including the second-order corrections as in (3.43):

$$H = -t\sum_{\langle i,j\rangle} c_i^{\dagger} c_j + g\omega_0 \sum_i \hat{n}_i (d_i + d_i^{\dagger}) + \sum_{\mathbf{q}} \omega_{\mathbf{q}} (d_{\mathbf{q}}^{\dagger} d_{\mathbf{q}} + 1/2), \tag{3.64}$$

where $d_{\mathbf{q}}$ is the Fourier transform of d_i. The phonon dispersion has been modeled using the intermolecular first neighbor force constant, $M\omega_1^2$, which yields, e.g., in 1D case $\omega_q^2 = \omega_0^2/2 + \omega_1^2 + (\omega_0^2/4 + \omega_0^2\omega_1^2 \cos q + \omega_1^2)^{1/2}$. MLF improves the convergence of the $1/\lambda$ perturbation series by introducing a suitable variational parameter $\lambda_{\mathbf{q}}$ in the LF transformation (3.35) as [144,145]

$$S \Rightarrow S_{\mathrm{MLF}} = \sum_{\mathbf{q},\mathbf{k}} \lambda_{\mathbf{q}} c_{\mathbf{k}+\mathbf{q}}^{\dagger} c_{\mathbf{k}} (d_{\mathbf{q}} - d_{-\mathbf{q}}^{\dagger}). \tag{3.65}$$

As a result, the polaron mass converges to much lower values when the phonon dispersion is introduced, in particular in the adiabatic regime. There is a continuous mass enhancement whose abruptness is significantly smoothed for the largest values of the phonon dispersion, similar to the SFP discussed above.

While a phase transition is ruled out in the single-electron Holstein Hamiltonian, where the ground-state energy is analytic in the EPI strength [36,38], a crossover from more itinerant to more self-trapped behavior may be identified as a rather sudden event in the adiabatic regime [146–149].

3.4 All-Coupling Lattice Polaron

3.4.1 Holstein Model at Any Coupling

During past 20 years, significant efforts were directed toward the extension of the weak- and strong-coupling perturbation lattice polaron theories to the intermediate region of the relevant parameters, $\lambda \sim 1$, and $\omega_0/t \sim 1$. It was argued [29,131,137] that the expansion parameter is actually $1/2z\lambda^2$, so the analytical strong-coupling expansion in powers of $1/\lambda$ might have a wider region of applicability than one can expect using simple physical arguments (i.e., $\lambda > 1$). However, it has not been clear how fast the expansion converges.

Kudinov and Firsov [138] developed the analytical approach to the two-site Holstein model by the use of the expansion technique, which provides the electronic and vibronic terms as well as the wave functions and all correlation functions in any order of powers of t. They have found the exponential reduction factor in *all* orders of the $1/\lambda$ perturbation expansion. On the other hand, the corrections to the atomic level were found as small as $1/\lambda^2$ rather than exponential in agreement with the conventional second-order result (3.43). Chatterjee and Das [144] studied the same problem for any coupling within the perturbative expansion combined with MLF (3.65) and MLF with a *squeezing* canonical transformation [150], $\exp(\tilde{S})$, where $\tilde{S} = \alpha(d_i d_i - d_i^\dagger d_i^\dagger)$. Using two variational parameters introduced by MLF and squeezing transformations allows for very good convergence of the $1/\lambda$ perturbation series even in the near-adiabatic regime, $\omega_0/t \gtrsim 0.5$, where the conventional $1/\lambda$ expansion shows bad convergence. These studies also showed that the region of the parameters of the Holstein model, where neither weak- nor strong-coupling perturbation analytical methods are applicable, is rather narrow. A semianalytical approach to the solution of two coupled differential equations (3.8) and (3.9) of the Holstein model in the whole parameter space has been proposed in [151] based on the coherent-state expansion of $u(x), v(x)$. These authors obtained the recursive relations among the expansion coefficients, allowing for highly accurate numerical solutions, which agree well with those by MLF method in the weak- and strong-coupling regimes. The deviation from the MLF solution in the intermediate-coupling regime implies that MLF misses some higher-order correlation terms. A continued fraction analytical solution of the two-site Holstein model was derived by Capone and Ciuchi [152] as for a related model in quantum optics [153]. In practice, it also requires some truncation of the infinite phonon Hilbert space. Finally, all Green's functions for the two-site Holstein–Hubbard model were derived in terms of continued fractions [154].

Numerical results obtained by different methods actually show that the ground-state energy (about $-E_{\mathrm{p}}$) is not very sensitive to the parameters, while the effective mass and the bandwidth strongly depend on the polaron size, EPI range, and the adiabatic ratio, ω_0/t. Several methods exist for numerical simulations of lattice polarons. They include exact diagonalization (ED) [35, 126, 155–163], the global–local (GL) [164], and other advanced variational methods [165,166]: quantum Monte Carlo (QMC) algorithms [167–177], density matrix renormalization group (DMRG) [178–180], continuous-time QMC [37,60,61,181,182], and diagrammatic QMC [55, 56, 183]. The methods vary in accuracy and versatility, and, combined together, can provide all the polaron properties of interest in the entire space of model parameters. On the other hand, ED suffers from the necessary truncation of the phonon Hilbert space, especially at strong couplings and low phonon frequencies (even then, the total Hilbert space is huge, reducing the number of sites and leading to poor momentum resolution), DMRG cannot easily handle long-range interactions, diagrammatic QMC and ED are inconvenient in calculating the density of states, and path-integral CTQMC slows down at small frequencies [35, 58, 177, 184]. In numerical analysis of polaron models, a complex approach is needed where each method is employed to calculate what it does best.

Until recently, most numerical studies were performed on the Holstein model (i.e., with zero-range EPI). Reliable results for the intermediate region were obtained using ED of vibrating clusters [35, 126, 139, 156, 157, 160–163, 185].

Taking as a measure of the polaron kinetic energy the correlation function $t_{\mathrm{eff}} = \langle -t(c_1^\dagger c_2 + c_2^\dagger c_1)\rangle$ (here $c_{1,2}$ are annihilation operators on the "left" and "right" sites of the Holstein model), one might doubt of the Lang–Firsov approach [139, 185], since this correlation function is much larger than the small-polaron bandwidth. However, applying the $1/\lambda$ expansion up to the second order in t, one obtains the numerical t_{eff} very close to the perturbation t_{LF} in the strong-coupling regime, $\lambda > 1$ [136, 186]:

$$t_{\mathrm{eff}} \simeq t_{\mathrm{LF}} \equiv -t \exp\left(-\frac{\lambda t}{\omega}\right) - \frac{t}{\lambda}, \tag{3.66}$$

with $\lambda \equiv 2E_{\mathrm{p}}/t$. Here, only the first exponential term describes the true coherent tunneling, while the second term describes the correction to the middle of the polaron band owing to the virtual "back-forth" transitions to the neighboring site, as discussed above (Sect. 3.2.1).

The main contribution to t_{eff} comes from the second-order term lowering the middle of the band [131, 136–138], rather than from the polaron transport-related first term (see also [187, 188]). Comparing the analytical expression (3.66) with the numerically calculated t_{eff}, one confirms that the Holstein–Lang–Firsov approach is asymptotically exact both in the nonadiabatic and

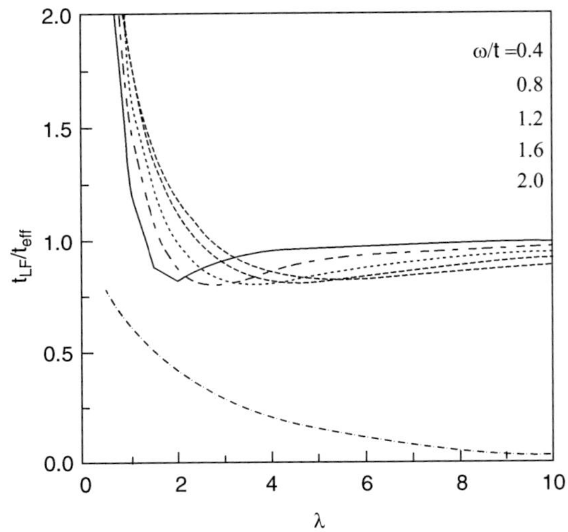

Fig. 3.3. The ratio of the perturbatively calculated correlator (3.66) of the two-site Holstein model to the numerically exact correlator for different values of the adiabatic ratio (here $\omega \equiv \omega_0$) [136]. The *lowest curve* represents the erroneous result of [185] with the missing second-order term in (3.66)

adiabatic regimes [136], if the second-order correction is taken into account (Fig. 3.3).

The numerical diagonalization of the two-site one-electron Holstein model [126] shows that the first-order term of the $1/\lambda$ perturbation theory describes well the polaron bandwidth in the *nonadiabatic* regime *for all values* of the coupling constant. There is no agreement in the *adiabatic* region, where the first-order perturbation expression (3.22) *overestimates* the polaron mass by a few orders of magnitude. A poor convergence of the perturbation expansion is due to appearance of the familiar double-well potential [18] in the adiabatic limit. The tunneling probability is extremely sensitive to the shape of this potential. The splitting of levels for the two-site cluster is well described by the Holstein quasiclassical formula generalized for the intermediate coupling in (3.33). While SHP is only a few times heavier than the bare (unrenormalized) electron in a wide range of the coupling for a moderate adiabatic ratio $\omega_0/t \gtrsim 1$, it becomes very heavy in the adiabatic regime and for the strong coupling [126].

3.4.2 Short-Range EPI in Infinite Lattices

A number of other independent numerical results proved that "by the use of the Holstein approximation and the canonical Lang–Firsov approach with appropriate corrections, one obtains an excellent estimate of the coherent bandwidth in both adiabatic and nonadiabatic (strong-coupling) regimes" [189].

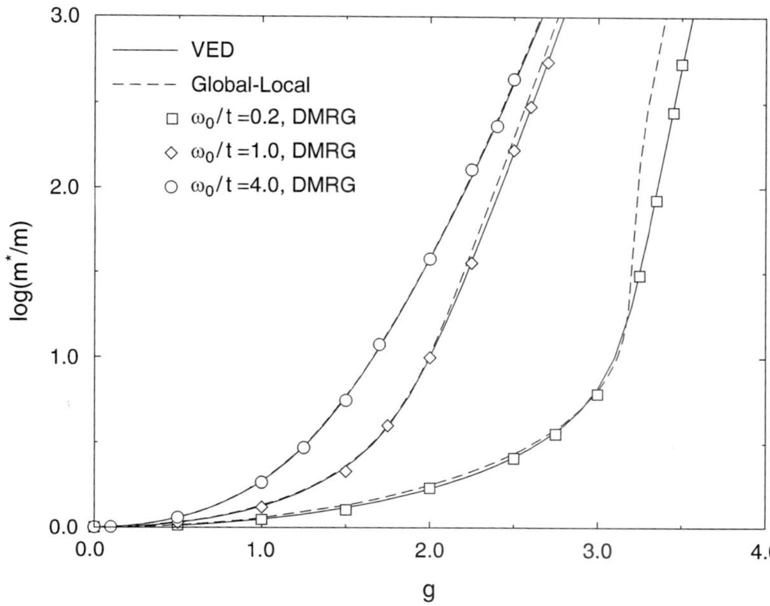

Fig. 3.4. Logarithm of the polaron effective mass in 1D Holstein model as a function of g (after [35]). VED results (*full lines*) were obtained operating repeatedly $L = 20$ times with the off-diagonal pieces of the Holstein Hamiltonian. For comparison, global–local (GL) results (*dashed lines*) are included [164]. *Open symbols*, indicating the value of ω_0/t, are DMRG results [178]

In particular, the elaborate variational ED (VED) [35,190] provides an exact numerical solution of the Holstein crystal model (3.64) in any dimension. In contrast to finite-lattice ED, it yields the ground-state energy which is a variational bound for the exact energy in the thermodynamic limit. Figure 3.4 shows the effective mass computed by VED [190] in comparison with GL and DMRG methods. m^* is obtained from $m/m^* = (1/2t)\partial^2 E_k/\partial k^2$, $k \to 0$, where $m_0 = 1/(2t)$ is the rigid band mass (here the lattice constant is $a = 1$). In the intermediate-coupling regime where VED gives an energy accuracy of 12 decimal places, one can calculate the effective mass accurately (6–8 decimal places) [35].

In Fig. 3.4, the parameters span different physical regimes including weak and strong coupling, and low and high phonon frequency. There is good agreement between VED and GL away from strong coupling and excellent agreement in all regimes with DMRG results. DMRG calculations are not based on finite-k calculations due to a lack of periodic boundary conditions, so they extrapolate the effective mass from the ground-state data using chains of different sizes, which leads to larger error bars and demands more computational effort. There is no phase transition in the ground state of the model, but the polaron becomes extremely heavy in the strong-coupling regime. The crossover to a regime of large polaron mass is more rapid in adiabatic regime, i.e., at small ω_0/t.

For analyzing all-coupling polarons in more complex lattices, the continuous-time path-integral quantum Monte Carlo algorithm (CTQMC) is ideally suited. The algorithm is formulated in real space [181, 184] and based on the analytical integration over phonon degrees of freedom introduced by Feynman [12] and on an earlier numerical implementation in discrete time by De Raedt and Lagendijk [168–171]. CTQMC introduced two critical improvements. Firstly, formulation in continuous imaginary time eliminated errors caused by the Trotter slicing and made the method numerically exact for any strength of EPI. Secondly, introduction of twisted boundary conditions in imaginary time [181, 191] enabled calculation of polaron effective masses, spectra, and the densities of states (DOS) in any dimensions in infinite lattices.

The polaron action, obtained by analytical integration over phonon degrees of freedom, is a functional of the polaron path in imaginary time $\mathbf{r}(\tau)$. It is given by the following double integral:

$$A[\mathbf{r}(\tau)] = \frac{z\lambda\bar{\omega}}{2\Phi_0(0,0)} \int_0^{\bar{\beta}} \int_0^{\bar{\beta}} \mathrm{d}\tau \mathrm{d}\tau' \mathrm{e}^{-\bar{\omega}\bar{\beta}/2} \left(\mathrm{e}^{\bar{\omega}(\bar{\beta}/2-|\tau-\tau'|)} + \mathrm{e}^{-\bar{\omega}(\bar{\beta}/2-|\tau-\tau'|)} \right)$$
$$\times \Phi_0[\mathbf{r}(\tau), \mathbf{r}(\tau')] + \frac{z\lambda\bar{\omega}}{\Phi_0(0,0)} \int_0^{\bar{\beta}} \int_0^{\bar{\beta}} \mathrm{d}\tau \mathrm{d}\tau' \mathrm{e}^{-\bar{\omega}\tau} \mathrm{e}^{-\bar{\omega}(\bar{\beta}-\tau')}$$
$$\times \left(\Phi_{\Delta\mathbf{r}}[\mathbf{r}(\tau), \mathbf{r}(\tau')] - \Phi_0[\mathbf{r}(\tau), \mathbf{r}(\tau')] \right), \tag{3.67}$$

$$\Phi_{\Delta\mathbf{r}}[\mathbf{r}(\tau), \mathbf{r}(\tau')] = \sum_{\mathbf{m}} \bar{f}_{\mathbf{m}}[\mathbf{r}(\tau)] \bar{f}_{\mathbf{m}+\Delta\mathbf{r}}[\mathbf{r}(\tau')], \tag{3.68}$$

where the vector $\Delta\mathbf{r} = \mathbf{r}(\bar{\beta}) - \mathbf{r}(0)$ is the difference between the end points of the polaron path, $\bar{\beta} = t/T$, $\bar{\omega} = \omega_0/t$, and $\bar{f}_{\mathbf{m}}(\mathbf{n}) = f_{\mathbf{m}}(\mathbf{n})/\kappa$ (see also (3.63)). From this starting point, the polaron is simulated using the Metropolis Monte Carlo method. The electron path is continuous in time with hopping events (or kinks) introduced or removed from the path with each Monte Carlo step. From this ensemble, various physical properties may be computed, in particular, the ground-state energy, the number of phonons in the polaron "cloud," and the polaron band energy spectrum:

$$\epsilon_{\mathbf{k}} - \epsilon_0 = - \lim_{\bar{\beta}\to\infty} \frac{1}{\bar{\beta}} \ln\langle \cos(\mathbf{k}\cdot\Delta\mathbf{r})\rangle, \tag{3.69}$$

where $\mathbf{k}$ is the quasimomentum. By expanding this expression in small $\mathbf{k}$, the ith component of the inverse effective mass is obtained as

$$\frac{1}{m_i^*} = \lim_{\bar{\beta}\to\infty} \frac{1}{\bar{\beta}} \langle (\Delta\mathbf{r}_i)^2 \rangle. \tag{3.70}$$

Thus, the inverse effective mass is the diffusion coefficient of the polaron path in the limit of the infinitely long "diffusion time" $\bar{\beta}$. The ground-state properties of the Holstein polaron are presented in Figs. 3.5–3.7.

3.4.3 Finite-Range EPI: Mobile Small Polaron

As discussed above (Sect. 3.3) the lattice polaron mass strongly depends on the radius of EPI. Also does the range of the applicability of the analytical $1/\lambda$ expansion theory. The theory appears almost exact in a wide region of the Fröhlich EPI (3.63) for which the exact polaron mass was calculated with CTQMC algorithm in [60].

At large λ (>1.5) SFP was found to be much lighter than SHP in agreement with the analytical results (Sect. 3.3), while the large Fröhlich polaron (i.e., at $\lambda < 1$) was *heavier* than the large Holstein polaron with the same binding energy (Fig. 3.8). The mass ratio $m^*_{\mathrm{FP}}/m^*_{\mathrm{HP}}$ is a nonmonotonic function of λ. The effective mass of the Fröhlich polaron, $m^*_{\mathrm{FP}}(\lambda)$ is well fitted by a single exponent which is $\mathrm{e}^{0.73\lambda}$ for $\omega_0 = t$ and $\mathrm{e}^{1.4\lambda}$ for $\omega_0 = 0.5t$. The exponents are remarkably close to those obtained with the Lang–Firsov transformation, $\mathrm{e}^{0.78\lambda}$ and $\mathrm{e}^{1.56\lambda}$, respectively. Hence, in the case of the Fröhlich interaction, the transformation is perfectly accurate even in the moderate adiabatic regime, $\omega_0/t \leq 1$ for *any* coupling strength. It is not the case for the Holstein polaron. If the interaction is short-ranged, the same analytical technique is applied only in the nonadiabatic regime $\omega_0/t > 1$ (see Fig. 3.6).

An important question about polaron properties also involves the effects of screening on the electron–phonon interaction. Unscreened EPI makes polarons very mobile [59], which leads to strong effects even on the qualitative physical properties of the polaron gas. Modeling the screening effects a form for the interaction force between electrons and phonons was introduced in [61], and is a screened discrete Fröhlich interaction:

$$f_{\mathbf{m}}(\mathbf{n}) = \frac{\kappa}{[(\mathbf{m}-\mathbf{n})^2+1]^{3/2}} \exp\left(-\frac{|\mathbf{m}-\mathbf{n}|}{R_{\mathrm{sc}}}\right). \tag{3.71}$$

It describes EPI of holes with c-axis polarized lattice distortions, which has been suggested as the relevant electron–phonon interaction in the cuprates [59].

The CTQMC polaron mass for the one-dimensional lattice with the screened Fröhlich EPI (3.71) is shown in Fig. 3.9 at four different values of the screening length, $R_{\mathrm{sc}} \to 0$ (the short-range Holstein interaction), $R_{\mathrm{sc}} = 1$, $R_{\mathrm{sc}} = 3$, and $R_{\mathrm{sc}} \to \infty$ (the nonscreened Fröhlich interaction).

The QMC results for the energy, and the number of phonons in the polaron cloud [61] are in excellent agreement with the weak-coupling expansion results at small λ, and tend to the strong-coupling result as $\lambda \to \infty$. An important observation is evident from the plot of $\ln(m^*/m_0)$ against λ shown in Fig. 3.9. At intermediate and large couplings (i.e., in the transition and small-polaron regions), altering the value of R_{sc} has a *dramatic* effect on the effective mass. For example, the nonscreened Fröhlich polaron is over 10^3 times "lighter" than the Holstein polaron at $\lambda = 4$, and over 10^4 times lighter at $\lambda = 5$. It

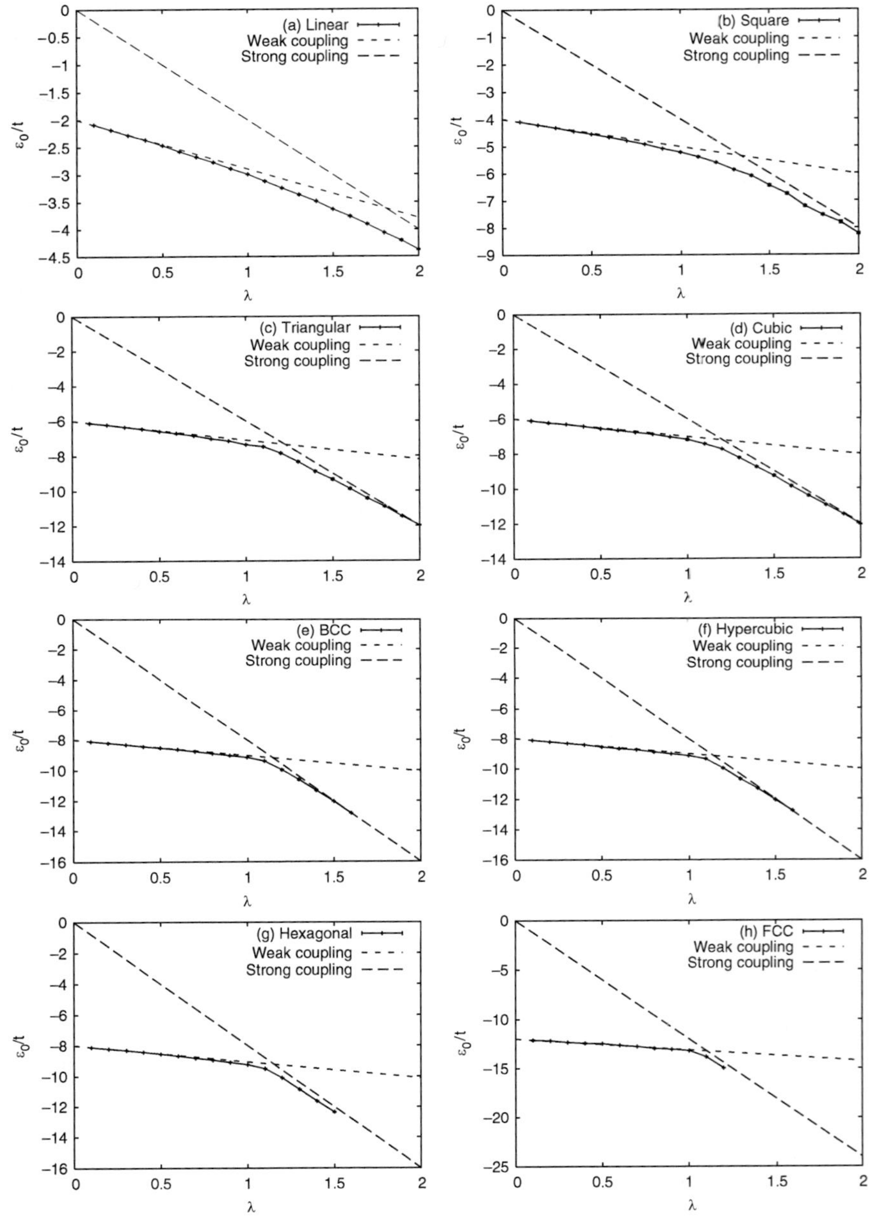

Fig. 3.5. Total energy of the Holstein polaron for increasing coupling λ with $\bar{\omega} = 1$ (after [37]). Note that the crossover from weak to strong coupling behavior is very fast for this quantity on the triangular and cubic lattices, but is much slower for the square lattice. In general, the speed of the crossover is quicker for larger coordination number. The curve lies below the strong-coupling asymptote for variational reasons and also lies below the weak-coupling result

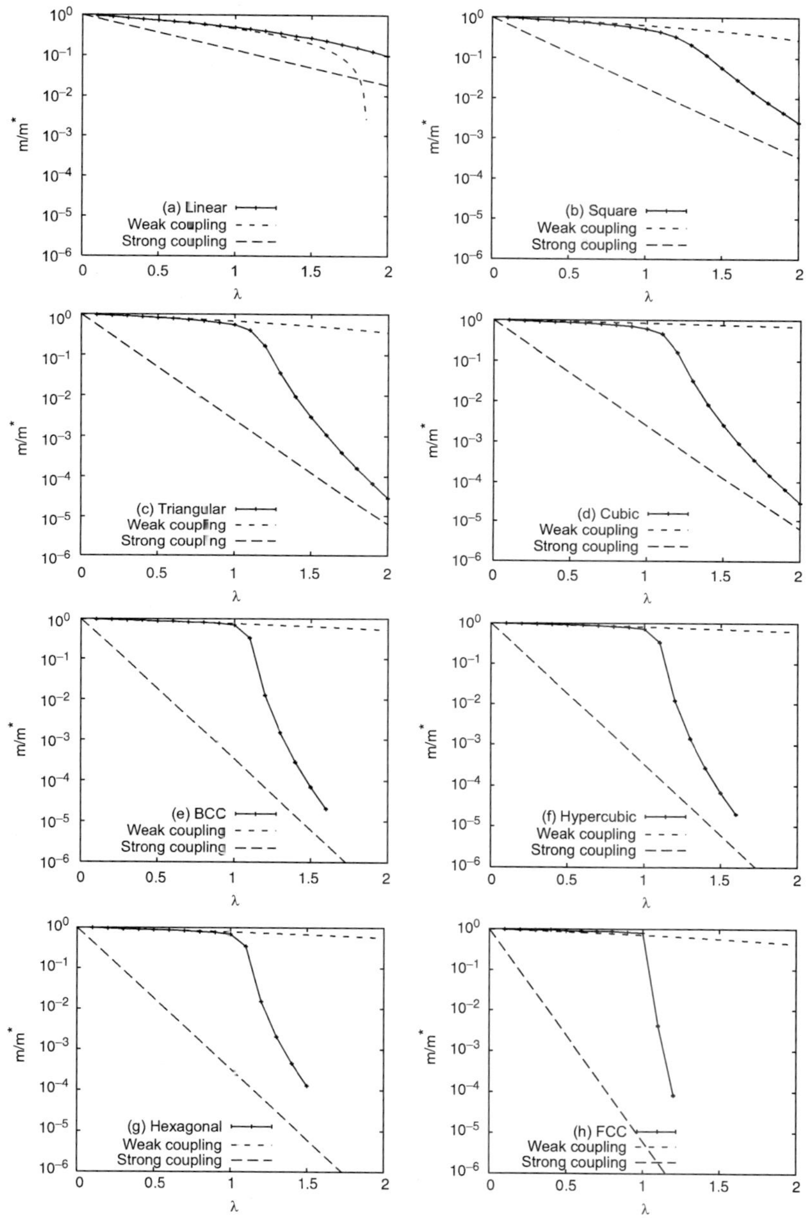

Fig. 3.6. Inverse effective mass for the Holstein polaron as a function of coupling λ on (**a**) linear, (**b**) square, (**c**) triangular, (**d**) cubic, (**e**) body-center-cubic, (**f**) hypercubic ($d = 4$), (**g**) hexagonal, and (**h**) face-center-cubic lattices, and $\bar{\omega} = 1$ (after [37]). Also shown are the weak- and strong-coupling asymptotes. Increased coordination number leads to more mobile polarons at weak coupling and more localized polarons for strong coupling with a very fast crossover between the two behaviors. *Error bars* show the standard error of the points

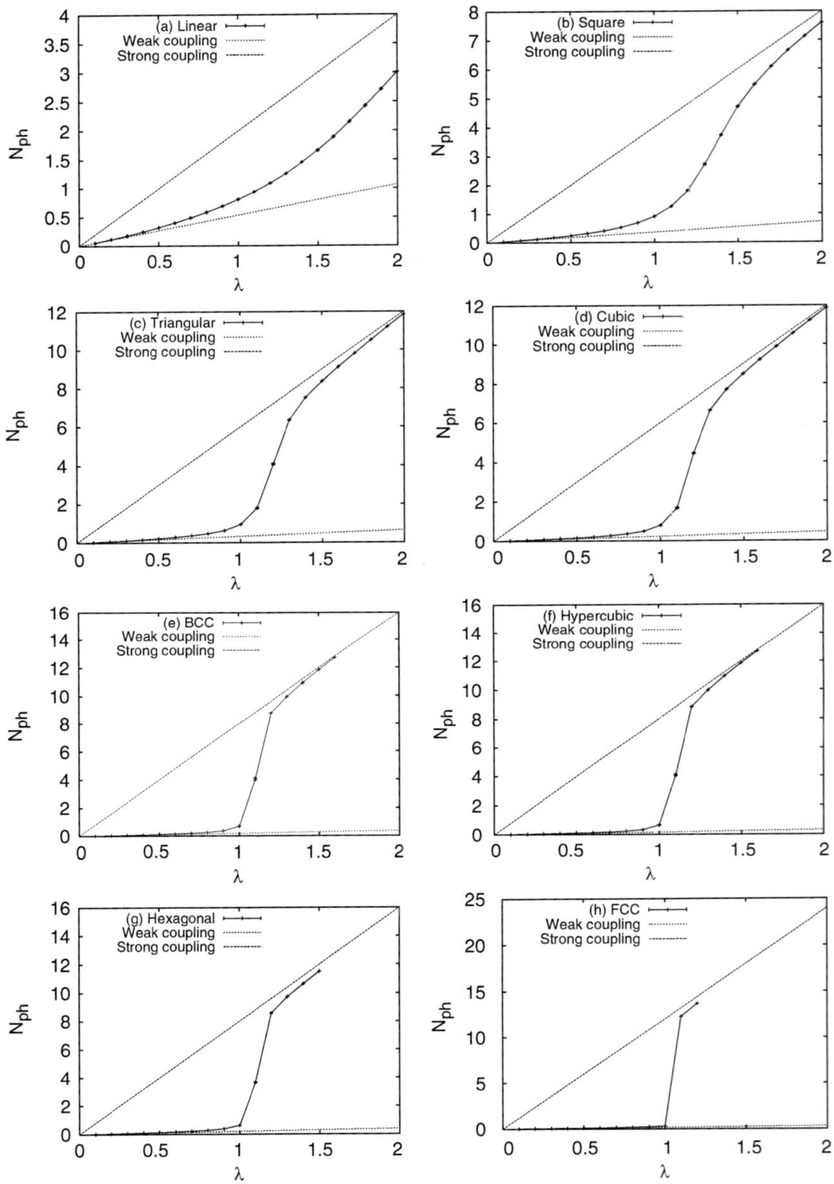

Fig. 3.7. Number of phonons associated with the Holstein polaron for increasing coupling with $\bar{\omega} = 1$ (after [37]). For most couplings, the number of phonons is closely related to the exponent of the effective mass. This is expected in the very strong-coupling limit where $m/m^* = \exp(-\gamma N_{\rm ph})$, and the continuation of this behavior to lower λ values is of interest

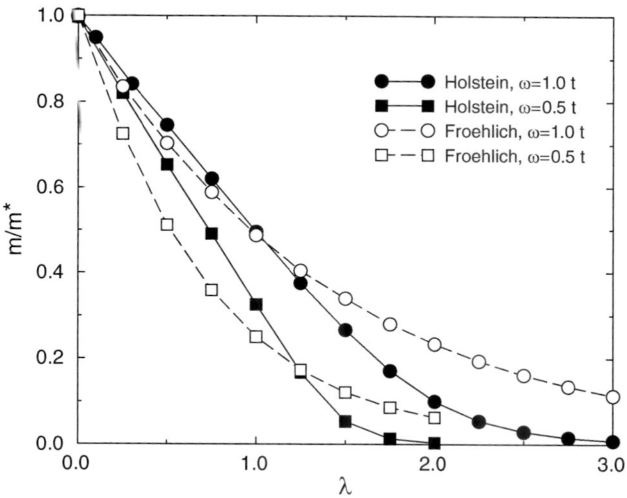

Fig. 3.8. Inverse effective polaron mass in units of the bare band mass in a linear lattice [60]

is apparent from the above results for the screened Fröhlich model that as R_{sc} increases (from Holstein to Fröhlich), the QMC results move, in general, closer to the "$1/\lambda$" expansion over the *entire* range of λ. That is, the "$1/\lambda$" expansion becomes generally more applicable as the range of EPI increases (as well as with increasing $\bar{\omega}$).

The polaron features for the long-range EPI were also investigated by extending a variational approach previously proposed for the study of systems with local (Holstein) coupling [192]. The ground-state spectral weight, the average kinetic energy, the mean number of phonons, and the electron–lattice correlation function were calculated for a wide range of model parameters focusing on the adiabatic regime. A strong mixing of electronic and phononic degrees of freedom even for small values of EPI was found in the adiabatic case due to the long-range interaction.

CTQMC has enabled accurate analysis of models with long-range electron–phonon interactions [37, 60, 61] and a model with anisotropic electron hopping [182].

All numerical results confirm gross polaronic features well understood analytically by Holstein [18] and others both in the nonadiabatic and adiabatic regimes. A great power of numerical methods is the ability to calculate an entire polaron spectrum, and the polaron DOS $\rho(E) = N^{-1}\sum_{\mathbf{k}} \delta(E - E_{\mathbf{k}} + E_0)$, in the whole parameter space. The coherent part of the spectrum, $\epsilon_{\mathbf{k}}$, possesses an interesting property of flattening at large lattice momenta in the adiabatic limit, $t \gg \omega_0$ [35, 162–164, 193, 194]. In the weak-coupling limit, the flattening can be readily understood as hybridization between the bare-electron spectrum and a phonon mode [195]. The resulting polaron dispersion is cosine-like at small $\mathbf{k}$ and flat at large $\mathbf{k}$. As a result, the polaron DOS

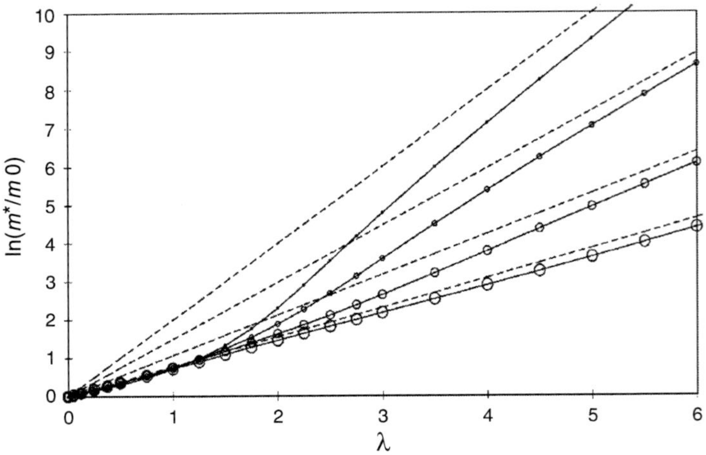

Fig. 3.9. The logarithm of the effective mass for screening lengths $R_{\mathrm{sc}} = 0, 1, 3, \infty$ (increasing size of circles) vs. λ at $\bar{\omega} = 1$ [61]. At intermediate and strong coupling, decreasing the value of R_{sc} dramatically increases the effective mass. The curves tend to the strong-coupling analytical result (*dashed lines*)

should be peaked close to the *top* of the polaron band. Exact VED [35] and CTQMC [184] calculations have confirmed that this is indeed the case for the short-range EPI. The evolution of the DOS of the isotropic Holstein model with phonon frequency ω_0 in two and three dimensions is shown in Fig. 3.10a,b. In all the cases presented the polaron is fully developed, with the bandwidth much smaller than ω_0.

At small ω_0, DOS develops a massive peak at the top of the band. The peak is more pronounced in $d = 3$ than in $d = 2$. The van Hove singularities are absorbed in the peak and as such cannot be seen. With increasing ω_0, the polaron spectrum approached the cosine-like shape in full accordance with the Lang–Firsov nonadiabatic formula. The respective DOS gradually assumes the familiar shape of the tight-binding band with renormalized hopping integrals. The van Hove singularities are clearly visible. These results have an interesting corollary for the Holstein model. At small-to-moderate ω_0/t in two and three dimensions, the bottom half of the polaron band contains a tiny minority of the total number of states, so that the system's responses will be dominated by the states in the peak.

In the long-range model (3.63) the two-dimensional DOS, shown in Fig. 3.10c, is much closer to the tight-binding shape than the Holstein DOS at the same parameters. The polaron spectrum and DOS are another manifestation of the extremity of the Holstein model. Long-range EPI removes those peculiarities and make the shape of polaron bands more close to the $1/\lambda$ expansion results.

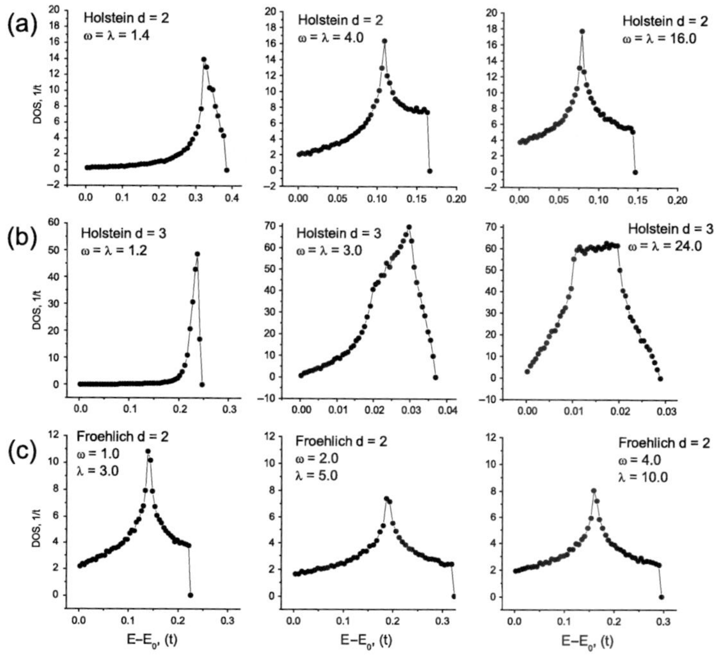

Fig. 3.10. (**a**) and (**b**) The evolution of the density of states of the Holstein polaron with phonon frequency $\omega = \omega_0/t$ in $d = 2$ and 3, respectively. (**c**) The same for the small Fröhlich polaron in $d = 2$ (after [184]

3.5 Isotope Effect on the Polaron Mass

There is a qualitative difference between ordinary metals and polaronic (semi) conductors. The renormalized effective mass of electrons is independent of the ion mass M in ordinary metals (where the Migdal adiabatic approximation is believed to be valid, see below), because λ does not depend on the isotope mass. However, when electrons form polarons dressed by lattice distortions, their effective mass m^* depends on M through $m^* = m \exp(\gamma E_{\mathrm{p}}/\omega_0)$, in the strong-coupling limit. Here, the phonon frequency depends on the ion mass, so that there is a large polaronic isotope effect (PIE) on the carrier mass with the carrier mass isotope exponent $\alpha_{m^*} = (1/2)\ln(m^*/m)$ as predicted in [196], in contrast to the zero isotope effect in ordinary metals (see [9] for more details). The effect was found experimentally in cuprates [197–199] and manganites [200]. More recent high-resolution angle-resolved photoemission spectroscopy (ARPES) also revealed a complicated isotope effect on the whole band structure in cuprate superconductors depending on the electron energy and momentum [201].

PIE in the intermediate region of parameters was calculated using the dynamic mean-field approximation (DMFT) [202, 203] and CTQMC algorithm [37, 61, 204]. Importantly, α_{m^*} and the effective mass averaged over dimensions are related to the critical temperature isotope exponent, $\alpha = -\mathrm{d}\ln T_{\mathrm{c}}/\mathrm{d}\ln M$, of a (bi)polaronic superconductor as

$$\alpha = \alpha_{m^*}\left(1 - \frac{m/m^*}{\lambda - \mu_{\mathrm{c}}}\right), \tag{3.72}$$

where μ_{c} is the Coulomb pseudopotential [9, 196].

In the adiabatic regime, the isotope effect on the polaron mass does not fully represent the isotope effect on the vast majority of polaron states, in particular for the short-range EPI, so that additional insight can be gained from the isotope effect on the entire polaron spectrum [204]. The isotope effect on polaron spectrum and DOS in $d = 2$ is illustrated in Fig. 3.11. The ratio of the two phonon frequencies, $\omega = 0.80t$ and $0.75t$, has been chosen to roughly correspond to the substitution of ^{16}O for ^{18}O in complex oxides. One can see that the polaron band shrinks significantly, by 20–30%, for both polaron types. The middle panels show the isotope exponents on spectrum points calculated as

$$\alpha_{\mathbf{k}} = \frac{1}{2}\frac{\langle\omega\rangle}{\langle\epsilon_{\mathbf{k}}\rangle}\frac{\Delta\epsilon_{\mathbf{k}}}{\Delta\omega}, \tag{3.73}$$

where the angular brackets denote the mean value either of the two frequencies or of the two energy values. An interesting observation is that $\alpha_{\mathbf{k}}$ of the Fröhlich polaron is roughly independent of $\mathbf{k}$ ($\pm 10\%$). In the Holstein case, $\alpha_{\mathbf{k}}$ dips in the vicinity of the Γ point.

Unconventional isotope effects as observed in high-temperature superconducting cuprates [197] were also explained by polaron formation stemming from the coupling to the particular quadrupolar $Q(2)$-type phonon mode [205]. PIEs in the spectral function of strongly correlated systems were numerically studied in [206] in the framework of the Holstein–Hubbard model and in [207] using the extended $t-J$ model including EPI.

3.6 Jahn–Teller Polaron

The density–displacement EPIs discussed above are not only possible types of the electron–lattice coupling. Examples of other types are the Su–Schrieffer–Heeger (SSH) EPI [208] (for recent path-integral results for SSH polarons, see [209]), and the Jahn–Teller (JT) EPI [210]. In the former, the lattice deformation is coupled to electron kinetic energy, while JT involves a multidimensional electron basis and a multidimensional representation of the deformation group. The JT interaction is active in some molecules and crystals of high point symmetry, and it was served as a guide in the search for high-temperature superconductivity [211]. Later on JT-type EPIs were widely discussed in connection with cuprate and other high-T_{c} superconductors (see, e.g., [212–216]) and colossal-magnetoresistance manganites (see, e.g., [217–221]).

The simplest model of the JT interaction is the $E \otimes e$ interaction [222] that describes a short-range coupling between twice-degenerate e_{g} electronic levels

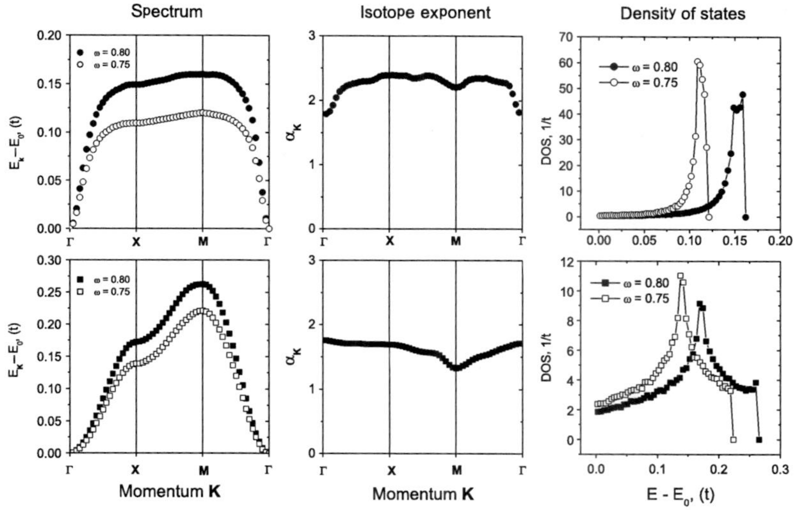

Fig. 3.11. Isotope effect on polaron spectrum and density of states. *Top row*: the 2*d* Holstein polaron at $\lambda = 1.2$. *Left*: polaron spectrum at $\omega = 0.80$ and 0.75 (in units of t). *Middle*: the isotope exponent for each **k**-point. *Right*: the density of states for the two frequencies. *Bottom row*: the same for the small 2*d* Fröhlich polaron at $\lambda = 2.4$ (after [204])

(c_1, c_2) and a local double-degenerate vibron mode (ζ, η). The Hamiltonian reads

$$\begin{aligned} H_{\mathrm{JT}} = & -t \sum_{\langle \mathbf{nn}' \rangle} \left(c^{\dagger}_{\mathbf{n}'1} c_{\mathbf{n}1} + c^{\dagger}_{\mathbf{n}'2} c_{\mathbf{n}2} \right) \\ & - \kappa \sum_{\mathbf{n}} \left[\left(c^{\dagger}_{\mathbf{n}1} c_{\mathbf{n}2} + c^{\dagger}_{\mathbf{n}2} c_{\mathbf{n}1} \right) \eta_{\mathbf{n}} + \left(c^{\dagger}_{\mathbf{n}1} c_{\mathbf{n}1} - c^{\dagger}_{\mathbf{n}2} c_{\mathbf{n}2} \right) \zeta_{\mathbf{n}} \right] \\ & + \sum_{\mathbf{n}} \left[-\frac{1}{2M} \left(\frac{\partial^2}{\partial \zeta^2_{\mathbf{n}}} + \frac{\partial^2}{\partial \eta^2_{\mathbf{n}}} \right) + \frac{M\omega^2}{2} \left(\zeta^2_{\mathbf{n}} + \eta^2_{\mathbf{n}} \right) \right] . \end{aligned} \tag{3.74}$$

The symmetry of the interaction ensures the same coupling parameter κ for the two phonons. Because the ionic coordinates of different cells are not coupled, the model describes a collection of separate clusters that are linked only by electron hopping. To relate the Hamiltonian to more realistic situations, phonon dispersion should be added [215, 223].

An important property of the $E \otimes e$ interaction is the absence of an exact analytical solution in the atomic limit $t = 0$. Here, in contrast with the Holstein and Frhlich EPI, the atomic limit is described by two coupled *partial* differential equations for the electron doublet $\psi_{1,2}(\zeta, \eta)$. At large couplings, however, the elastic energy assumes the Mexican hat shape and the phonon dynamics separates into radial oscillatory motion and azimuthal rotary motion. This results in an additional pre-exponential factor $\propto \kappa$ in the ion overlap integral, leading to the effective mass $m^*_{\mathrm{JT}}/m = (2/\pi g)^{1/2} \exp g^2$, where $g^2 = \kappa^2/2M\omega^3$ [224].

A path-integral approach to Hamiltonian (3.74) was developed by Kornilovitch [225]. Its details are found in [184]. Because there are two electron orbitals, the electron path must be assigned an additional orbital index (or *color*) $a = 1, 2$. Color 1 (or 2) of a given path segment means that it resides in the first (second) atomic orbital of the electron doublet. There is a difference between the two phonons. Phonon ζ is coupled to electron density, like in the Holstein case. The difference from the Holstein is that the direction of the force changes to the opposite when the electron changes orbitals. In contrast, phonon η is coupled to orbital changes themselves: the more often the electron changes orbitals, the more "active" is η. Discrete orbital changes are analogous to electron hops between discrete lattice sites, and as such are associated with "kinetic orbital energy." Phonon coupling to orbital changes is analogous to phonon coupling to electron hopping in the SSH model [208].

After the update rules are established, the JT polaron properties can be calculated with no approximations using QMC algorithm. The mass, spectrum, and density of states are obtained as for the conventional lattice polarons. Results of QMC calculations are shown in Fig. 3.12 [225]. Most

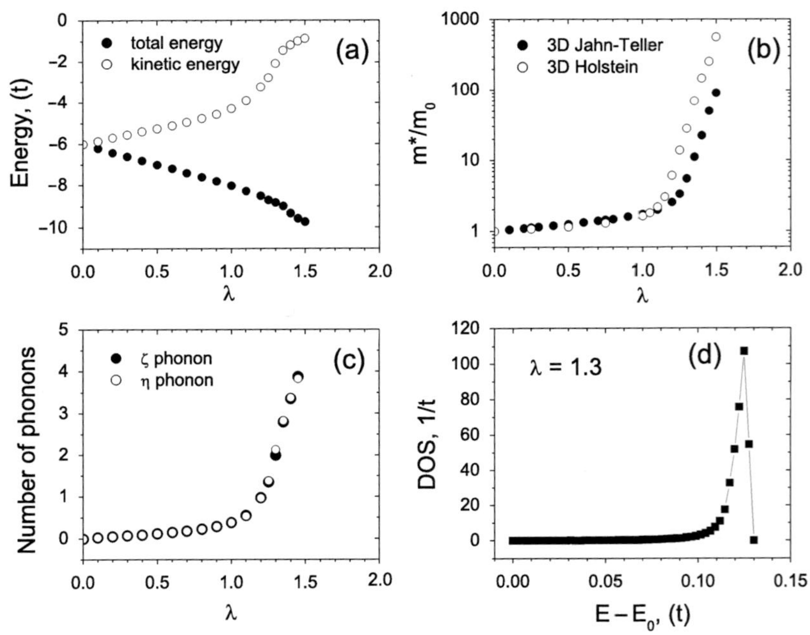

Fig. 3.12. Physical properties of the $d = 3$ Jahn–Teller polaron at $\omega = 1.0t$ [225]. (**a**) The total and kinetic energy. (**b**) The effective mass compared with the $d = 3$ Holstein polaron at the same phonon frequency. (**c**) The number of excited phonons of both types. (**d**) The density of states of the JT polaron at $\lambda = 1.3$ (after [184]

properties behave similarly to those of the $d = 3$ Holstein polaron at the same phonon frequency. For example, the kinetic energy (Fig. 3.12a) sharply decreases by absolute value between $\lambda = 1.2$ and 1.4. The JT polaron mass is slightly larger at the small to intermediate coupling, but several times smaller at the strong coupling. This nonmonotonic behavior of the ratio of the JT and Holstein masses was later confirmed by accurate variational calculations [226], although in that work the JT polaron (and bipolaron) was investigated in one spatial dimension. The relative lightness of the JT polaron is consistent with Takada's result mentioned above [224]. The number of excited phonons of both types is shown in Fig. 3.12c. Interestingly, the shape of the phonon curves is similar to that of the logarithm of the effective mass. This suggests an intimate relationship between the two quantities, again similarly to the Holstein case. Finally, the density of JT polaron states features the same peak at the top of the band, caused by the spectrum flattening at large polaron momenta.

In summary, the local character of the JT interaction and the independence of vibrating clusters result in the same extremity of polaron properties as in the 3*d* Holstein model. One could expect that either a long-range JT interaction or phonon dispersion will soften the sharp polaron features and make JT polarons more mobile.

3.7 Trapping of Lattice Polarons by Impurities

The situation as regards polaron formation and dynamics in real materials is complicated by an intrinsic disorder. In a pioneering paper, Economou and coauthors [227] studied a one-dimensional (1D) large polaron with a diagonal disorder by using methods from the theory of nonlinear systems. Bronold et al. [228] investigated the dynamics of a single electron in a Holstein model with a site-diagonal, binary-alloy-type disorder by applying a dynamical mean-field theory (DMFT) for a Bethe lattice with infinite coordination number.

Since a delicate interplay between the self-trapping by EPI and the trapping by doped-induced disorder in the intermediate-coupling regime, even a single polaron has to be studied by numerically exact CTQMC or DMC techniques.

Using CTQMC, Hague *et al.* [229] have solved the problem of the electron interacting with phonons in the presence of an impurity in a 3D lattice. The electron hopping was assumed to be between nearest neighbors only. The phonon subsystem was made up of independent oscillators with frequency ω, displacement $\xi_{\mathbf{m}}$, momentum $\hat{P}_{\mathbf{m}} = -i\hbar\partial/\partial\xi_{\mathbf{m}}$, and mass M associated with each lattice site. The sites are indexed by $\mathbf{n}$ or $\mathbf{m}$ for electrons and ions, respectively. The real-space Hamiltonian reads

$$H = -t \sum_{\langle \mathbf{nn}' \rangle} c^\dagger_{\mathbf{n}'} c_{\mathbf{n}} + \sum_{\mathbf{n}} \Delta_{\mathbf{n}} c^\dagger_{\mathbf{n}} c_{\mathbf{n}} + \frac{1}{2M} \sum_{\mathbf{m}} \hat{P}^2_{\mathbf{m}} + \frac{M\omega^2}{2} \sum_{\mathbf{m}} \xi^2_{\mathbf{m}} - \sum_{\mathbf{n}\mathbf{m}\sigma} f_{\mathbf{m}}(\mathbf{n}) c^\dagger_{\mathbf{n}} c_{\mathbf{n}} \xi_{\mathbf{m}}. \quad (3.75)$$

Here, $\langle \mathbf{nn}' \rangle$ denote pairs of nearest neighbors. The spin indices and Hubbard U are omitted since there is only one electron. As above, the dimensionless e–ph coupling constant λ is defined as $\lambda = \sum_{\mathbf{m}} f^2_{\mathbf{m}}(0)/2M\omega^2 zt$ which is the ratio of the polaron energy when $t = 0$ to the kinetic energy of the free electron $W = zt$. Hague *et al.* [229] simulated the trapping of the Holstein polaron, with the force function $f_{\mathbf{m}}(\mathbf{n}) = \delta_{\mathbf{n},\mathbf{m}}$ for the case where $\Delta_{\mathbf{n}} = \Delta\delta_{\mathbf{n},\mathbf{0}}$, which is the single impurity problem.

One of the main complications regarding the QMC simulation of a particle in a single impurity potential is ensuring that the whole configuration space is sampled. To ensure that the path includes sufficient sampling of the impurity, a functional of the path has been introduced with the property that the walker revisits the impurity in finite time. In the weighted ensemble, the ground-state polaron energy, the number of phonons in the polaron cloud, and the average distance from the impurity were computed.

Applying the Lang–Firsov canonical transformation and assuming large phonon frequency, an approximate instantaneous Hamiltonian can be derived

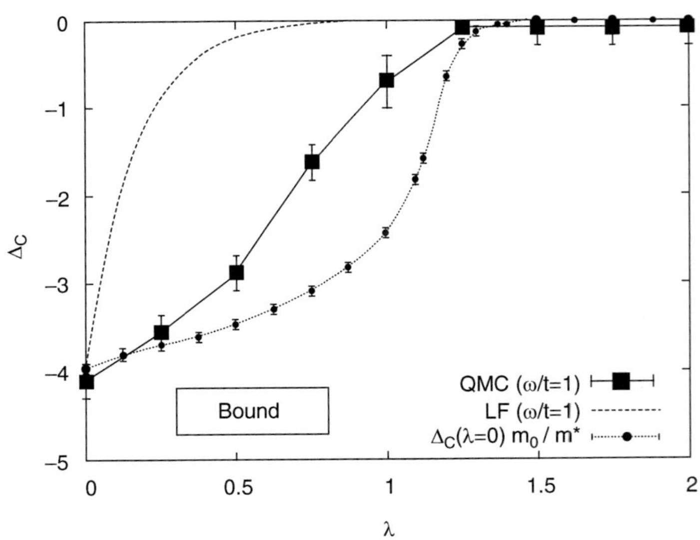

Fig. 3.13. Binding diagram in λ, Δ space. Also shown are the approximate values predicted by the Lang–Firsov transformation with zero excited phonons $\Delta_C(\lambda) = \Delta_C(0)\exp(-zt\lambda/\omega)$, from the exact mass in the perfect crystal lattice (measured using QMC) $\Delta_C(\lambda) = \Delta_C(0)m_0/m^*$ and the value measured from the Monte Carlo code (after [229])

with the form: $\tilde{H} = -\tilde{t}\sum_{\langle ij\rangle} c_i^\dagger c_j + \Delta n_0$, where $\tilde{t} = t\exp(-zt\lambda/\omega)$. It is possible to compute an approximate analytic value for the value of Δ at which impurity trapping occurs, $\Delta_\mathrm{C} = 3.95\tilde{t} = 3.95m_0/m^*$. When there is no electron–phonon coupling, the relation $\Delta_\mathrm{C} = 3.95t$ is exact. While the approximation for $\tilde{t}$ is not expected to hold for low phonon frequency, use of the exact value of the inverse mass computed from CTQMC is expected to lead to a qualitatively correct value for Δ_C.

Figure 3.13 shows the binding diagram in λ, Δ space. Also shown are the approximate values predicted by the Lang–Firsov transformation with zero excited phonons $\Delta_\mathrm{C}(\lambda) = \Delta_\mathrm{C}(0)\exp(-zt\lambda/\omega)$, from the exact mass in the noninteracting case (measured using QMC) $\Delta_\mathrm{C}(\lambda) = \Delta_\mathrm{C}(0)m_0/m^*$ and the value measured from the current Monte Carlo code (here m_0 is the bare band mass). The total energy showed that for electron–phonon coupling $\lambda > 1$, Holstein polarons are strongly bound to the impurity. For small λ, there is a more gradual binding, coupled with a sudden increase in the number of phonons present in the polaron. The binding diagram (Fig. 3.13) has showed that in the intermediate-coupling regime, the critical impurity strength changes monotonically with λ and differs significantly from the weak- and strong-coupling approximations.

3.8 Response of Lattice Polarons

3.8.1 Hopping Mobility

Studies of lattice polarons allowed for a theoretical understanding of low mobility problem [230] of many "poor" conductors, where an estimate of the mean-free path yields values much lower than the lattice constant. Transport properties of lattice polarons depend strongly on temperature. For T lower than the characteristic phonon energy polaron kinetics is the Boltzmann kinetics of heavy particles tunneling in the narrow band [18]. However at higher temperatures, the polaron band collapses (Sect. 3.2.2), and the transport is diffusive via thermally activated jumps of polarons from site to site [18, 22, 27, 29, 129, 231–233].

The $1/\lambda$ perturbation expansion is particularly instrumental in calculating hopping mobilities and optical absorption coefficients. Applying the canonical transformation (Sect. 3.2) and singling out the diagonal coherent tunneling in the polaron band, one can write the Hamiltonian as

$$\tilde{H} = H_\mathrm{p} + H_\mathrm{ph} + H_\mathrm{int}, \tag{3.76}$$

where $H_\mathrm{p} = \sum_\mathbf{k} \xi_\mathbf{k} c_\mathbf{k}^\dagger c_\mathbf{k}$ is the "free" *polaron* contribution (here we drop the spin), $H_\mathrm{ph} = \sum_\mathbf{q} \omega_\mathbf{q}(d_\mathbf{q}^\dagger d_\mathbf{q} + 1/2)$ is the phonon part, and $\xi_\mathbf{k} = Z'E_\mathbf{k} - \mu$ is the renormalized (polaron) band dispersion. The chemical potential μ includes the polaron level shift $-E_\mathrm{p}$, and it could also include all higher orders in $1/\lambda$ corrections to the polaron spectrum, which are independent of $\mathbf{k}$ (3.43). The band-narrowing factor Z' is defined as

$$Z' = \frac{\sum_{\mathbf{m}} t(\mathbf{m}) e^{-g^2(\mathbf{m})} \exp(-i\mathbf{k} \cdot \mathbf{m})}{\sum_{\mathbf{m}} t(\mathbf{m}) \exp(-i\mathbf{k} \cdot \mathbf{m})}, \quad (3.77)$$

which is $Z' = \exp(-\gamma E_p/\omega)$, where $\gamma \leq 1$ depends on the range of EPI and phonon frequency dispersions (Sect. 3.3). The interaction term H_{int} comprises the polaron–polaron interaction (3.40) and the residual *polaron–phonon* interaction:

$$H_{\text{p-ph}} \equiv \sum_{i \neq, j} [\hat{\sigma}_{ij} - \langle \hat{\sigma}_{ij} \rangle] c_i^\dagger c_j, \quad (3.78)$$

where $\langle \hat{\sigma}_{ij} \rangle$ means averaging with respect to the bare phonon distribution. In the framework of the single-polaron problem, one neglects the polaron–polaron interaction and treats $H_{\text{p-ph}}$ as a perturbation.

The motion of the small polaron at high temperatures is a random walk consisting of steps from site to site [18]. Holstein calculated the probability W for the hop of a small polaron to a neighboring site. He suggested that the random walk was a Markovian process. For such processes, the diffusion coefficient is given by $D \simeq a^2 W$, where W is the hopping probability. The only term in the polaronic Hamiltonian, which changes the phonon occupation numbers, is the polaron–phonon interaction $H_{\text{p-ph}}$. The nearest-neighbor hopping probability in the second order is

$$W = 2\pi \left\langle \sum_j |\langle j|H_{\text{p-ph}}|i\rangle|^2 \delta\left(\sum_{\mathbf{q}} \omega_{\mathbf{q}} (n_{\mathbf{q}}^j - n_{\mathbf{q}}^i)\right)\right\rangle, \quad (3.79)$$

where $|i\rangle$ and $|j\rangle$ are the eigenstates of H_0 corresponding to the polaron on site i with $n_{\mathbf{q}}^i$ phonons in each phonon mode and the polaron on the neighboring site j with $n_{\mathbf{q}}^j$ phonons, respectively. Replacing the δ-function in (3.79) by the integral yields

$$W = t^2 e^{-2g^2} \int_{-\infty}^{\infty} d\tau \left[\exp\left(\frac{1}{N} \sum_{\mathbf{q}} |\gamma(\mathbf{q})|^2 [1 - \cos(\mathbf{q} \cdot \mathbf{a})] \times \frac{\cos(\omega_{\mathbf{q}} \tau)}{\sinh(\omega_{\mathbf{q}}/2T)}\right) - 1\right] e^{-i0^+|\tau|}. \quad (3.80)$$

The integration over τ is performed using a saddle-point approximation by allowing a finite phonon frequency dispersion $\delta\omega \ll \omega_{\mathbf{q}}$. Expanding $\cos(\omega_{\mathbf{q}}\tau)$ in powers of τ, and using the Einstein relation $\mu = eD/T$, one obtains the hopping mobility:

$$\mu_{\text{h}} = ea^2 \frac{\pi^{1/2} t^2}{2T(E_a T)^{1/2}} e^{-E_a/T}, \quad (3.81)$$

where

$$E_a = \frac{T}{N} \sum_{\mathbf{q}} |\gamma(\mathbf{q})|^2 [1 - \cos(\mathbf{q} \cdot \mathbf{a})] \tanh(\omega_{\mathbf{q}}/4T) \quad (3.82)$$

is the activation energy, which is *half* of the polaron level shift E_p for the Holstein EPI. This expression is applicable at $T \gg \omega_0/2$ and $t^2/\omega_0(E_a T) \ll 1$. Importantly, increasing the EPI range diminishes the value of E_a further. The hopping mobility $\mu_h \equiv \sigma_h/ne \sim \exp(-E_a/T)$ can be below $ea^2 \simeq 1\,\mathrm{cm}^2\,\mathrm{Vs}^{-1}$, which is the lowest limit for the Boltzmann theory to be applied. Within the Boltzmann theory, such a low mobility corresponds to the mean free path $l < a$, which is not a reasonable result.

Holstein suggested that at low temperatures $T < \omega_0/2 \ln \gamma_0$, there is the ordinary Boltzmann transport in momentum space, but in the narrow polaron band. According to Holstein [18], the transition from the band regime to the hopping regime occurs when the uncertainty in the polaron energy becomes comparable to the width of the polaron band. If phonons dominate in the scattering, the polaron mobility, $\mu = \mu_t + \mu_h$ decreases when the temperature increases from zero to some $T_{\min}$, where it is at minimum, because its Boltzmann part, μ_t, falls down due to an increasing number of phonons, while the hopping part, μ_h, remains small. However, μ increases above $T_{\min}$ due to the thermal activated hopping. There is a wide temperature range around $T_{\min}$ where the thermal activated hopping still makes a small contribution to the conductivity, but the uncertainty in the polaron band is already significant [29, 234]. The polaron transport theory requires a special diagrammatic technique [29, 129, 232, 234] and a conditional-probability function description [235, 236] in this region.

The transverse conductivity σ_{xy} and the Hall coefficient $R_H = \sigma_{xy}/H\sigma_{xx}^2$ of lattice polarons can be calculated with the Peierls substitution [237]:

$$t(\mathbf{m}-\mathbf{n}) \rightarrow t(\mathbf{m}-\mathbf{n})\mathrm{e}^{-\mathrm{i}e\mathbf{A}(\mathbf{m})\cdot(\mathbf{m}-\mathbf{n})}, \tag{3.83}$$

which is a fair approximation if the magnetic field, B, is weak compared with the atomic field $eBa^2 \ll 1$. Here, $\mathbf{A}(\mathbf{r})$ is the vector potential, which can also be time dependent. Within the Boltzmann theory, the sign of the Hall coefficient $R_H \simeq \pm 1/en$ depends on the type of carriers (holes or electrons) and the Hall mobility $\mu_H \equiv R_H \sigma_{xx}$ is the same as the drift mobility μ_t up to a numerical factor of the order of 1. The calculations of the hopping Hall current similar to those of the hopping conductivity [238–241] show that the Hall mobility depends on the symmetry of the crystal lattice and has nothing in common with the hopping mobility, neither with respect to the temperature dependence and even nor with respect to the sign. In particular for hexagonal lattices, three-site hops yield [241]

$$\mu_H = ea^2 \frac{\pi^{1/2} t}{(12 E_a T)^{1/2}} \mathrm{e}^{-E_a/3T} \tag{3.84}$$

with the *same sign* for electrons and holes. The activation energy of the Hall mobility is three times less than that of the hopping mobility. In cubic crystals, the hopping Hall effect is governed by four-site hops. The four-site calculations in [239, 240] gave the Hall mobility with the "normal" sign depending on the type of carriers.

The Hall conductivity and resistivity of strongly localized electrons at low temperatures and small magnetic fields strongly depend on frequency, the size of a sample [242], and on a magnetic order like, for example, in ferromagnetic (Ga,Mn)As [243]. In the presence of the spin–orbit interaction, each hopping path acquires a spin-dependent phase factor of the same form as that in a perpendicular (to the 2D system) magnetic field, which leads to spin accumulation and spin-Hall effects [238, 244].

3.8.2 Optical Conductivity

One of the fingerprints of lattice polarons is the frequency (ω) and temperature dependence of their mid-infrared (MIR) conductivity $\sigma(\omega)$. In the low-frequency and low-temperature region, where the tunneling band transport operates [17, 18], the conductivity acquires the conventional Drude form

$$\sigma(\omega) = \frac{n_{\mathrm{p}} e \mu_{\mathrm{t}}}{1 + (\omega\tau)^2}, \tag{3.85}$$

where the transport relaxation time τ may be frequency dependent because of the narrow band. For high (mid-infrared) frequencies, well above the polaron bandwidth ($\omega \gg w$) but below the interband gap, the Drude law is not applied. In this frequency region, one can apply the generalized Einstein relation $\sigma(\omega) = eD(\omega)/\omega$, where $D(\omega) = a^2 W(\omega)$, and $W(\omega)$ is the hopping probability of the absorption of the energy quantum ω [18]. The number of nearest-neighbor transitions per second with the absorption of a photon of the energy ω is given by the Fermi–Dirac golden rule:

$$W^- = 2\pi \left\langle \sum_j |\langle i|H_{\mathrm{p\text{-}ph}}|j\rangle|^2 \delta\left(\sum_{\mathbf{q}} \omega_{\mathbf{q}}(n_{\mathbf{q}}^j - n_{\mathbf{q}}^i) + \omega\right)\right\rangle, \tag{3.86}$$

and with the emission,

$$W^+ = 2\pi \left\langle \sum_j |\langle i|H_{\mathrm{p\text{-}ph}}|j\rangle|^2 \delta\left(\sum_{\mathbf{q}} \omega_{\mathbf{q}}(n_{\mathbf{q}}^j - n_{\mathbf{q}}^i) - \omega\right)\right\rangle. \tag{3.87}$$

As a result, one obtains

$$W(\omega) = W^- - W^+ = 2t^2 \mathrm{e}^{-2g^2} \sinh(\omega/2T) \tag{3.88}$$

$$\times \int_{-\infty}^{\infty} \mathrm{d}\tau \mathrm{e}^{-\mathrm{i}\omega\tau} \left[\exp\left(\frac{1}{N}\sum_{\mathbf{q}} |\gamma(\mathbf{q})|^2 [1 - \cos(\mathbf{q}\cdot\mathbf{a})] \frac{\cos(\omega\tau)}{\sinh(\omega_{\mathbf{q}}/2T)}\right) - 1\right]. \tag{3.89}$$

As in the case of the hopping mobility discussed above, the integral over τ is calculated using the saddle-point approximation [28]:

$$\sigma(\omega) = \sigma_{\mathrm{h}} \frac{T \sinh(\omega/2T)}{\omega} \mathrm{e}^{\omega^2/4\delta}, \tag{3.90}$$

where

$$\delta = \frac{1}{2N}\sum_{\mathbf{q}} |\gamma(\mathbf{q})|^2[1-\cos(\mathbf{q}\cdot\mathbf{a})]\frac{\omega_{\mathbf{q}}^2}{\sinh(\omega_{\mathbf{q}}/2T)}.$$

For high temperatures, $T \gg \omega_0/2$, this expression simplifies as

$$\sigma(\omega) = n_{\mathrm{p}}e^2a^2\frac{\pi^{1/2}t^2[1-\mathrm{e}^{-\omega/T}]}{2\omega(E_aT)^{1/2}}\exp\left[-\frac{(\omega-4E_a)^2}{16E_aT}\right]. \quad (3.91)$$

Here, $\sigma_{\mathrm{h}} = ne\mu_{\mathrm{h}}$ and n_{p} (n) is the atomic density (concentration) of polarons. The frequency dependence of the MIR conductivity has a form of an asymmetric Gaussian peak centered at $\omega = 4E_a$ with the half-width $4\sqrt{E_aT}$ [16, 245, 246]. According to the Franck–Condon principle, the position of the ions is not changed during an optical transition. Therefore, the frequency dependence of the MIR conductivity can be understood in terms of transitions between the adiabatic levels of the two-site Holstein model (Sect. 3.1.2). The polaron, say, in the left potential well absorbs a photon through the vertical transition to the right well, where the deformation is lacking, without any change in the molecular configuration. The photon energy required to excite the polaron from the bottom of the well is $\omega \simeq 2E_{\mathrm{p}}$, which corresponds to the maximum of $\sigma(\omega)$ in (3.91). The main contribution to the absorption comes from the states near the bottom with the energy of the order of T. The corresponding photon energies are found in the interval $2E_{\mathrm{p}} \pm \sqrt{8E_{\mathrm{p}}T}$ in agreement with (3.91). For low temperatures $T < \omega_0/2$, the half-width of the MIR maximum is about $\sim\sqrt{E_{\mathrm{p}}\omega_0}$ rather than $\sim \sqrt{E_{\mathrm{p}}T}$ [247]. The optical absorption of small polarons is distinguished from that of large polarons (Sect. 2.4.6) by the shape and the temperature dependence. Their comparison [247–249] shows a more asymmetric and less temperature-dependent MIR absorption of large polarons compared with that of small polarons. The high-frequency behavior of the optical absorption of small polarons is described by a Gaussian decay [250] (3.91) while for large polarons it is much slower power law $\omega^{-5/2}$ [93].

Many contemporary materials, such as superconducting cuprates and colossal magnetoresistive manganites, are characterized by intermediate values of EPI, $\lambda \sim 1$, which requires an extension of the theory of optical absorption to the crossover region from continuous to lattice polarons. The intermediate-coupling and frequency regime has been inaccessible for an analytical or semianalytical analysis, with an exception of infinite spatial dimensions, where DMFT yields reliable results [251–254]. DMFT treats the local dynamics exactly, but it cannot account for the spatial correlations being important in real finite-dimensional systems. Nevertheless, DMFT allows one to address the intermediate-coupling and adiabaticity regimes in the absorption not covered by the standard small-polaron theory, where qualitatively new features arise. In particular, the optical absorption exhibits a reentrant behavior, switching from weak-coupling-like to polaronic-like upon increasing the temperature, and sharp peaks with a nonmonotonic temperature dependence emerge at characteristic phonon frequencies [255].

Earlier ED studies of the lattice polaron absorption were limited to small two- to ten-site 1D and 2D clusters in the Holstein model [156, 189, 256, 257]. The optical absorption occurs as energy is transferred between the electromagnetic field and the phonons via the charge carriers. The vibration energy must be capable of being dissipated. Hence, using ED one has to introduce some continuous density of phonon states, or a phonon lifetime, which makes MIR absorption to be finite [256]. As a result, one obtains a fair agreement between ED absorption spectra and the analytical results (3.91) in the strong-coupling limit as far as a smooth part of ω dependence is concerned. The MIR conductivity occurs much more asymmetric in the intermediate-coupling region than in the strong-coupling regime, and it shows an additional oscillating superstructure corresponding to a different spectral weight of the states with a different number of virtual phonons in the polaron cloud.

More recent ED, VED, and a kernel polynomial method (KPM) (for a review of KPM, see [258]) allowed for numerical calculations of lattice polaron properties in the Holstein model in the whole parameter range on fairly large systems [155,226,259]. Applying standard linear-response theory, the real part of the conductivity takes the form: $\mathrm{Re}\,\sigma(\omega) = \mathcal{D}\delta(\omega) + \sigma^{\mathrm{reg}}(\omega)$, where $\mathcal{D}$ denotes the Drude weight at $\omega = 0$ and σ^{reg} is the finite-frequency response for $\omega > 0$. It can be written in spectral representation at $T = 0$ as [7]

$$\sigma^{\mathrm{reg}}(\omega) = \frac{\pi}{\omega N} \sum_{E_m > E_0} |\langle\psi_m|\hat{\jmath}|\psi_0\rangle|^2 \; \delta[\omega - (E_m - E_0)] \tag{3.92}$$

with the (paramagnetic) current operator $\hat{\jmath} = -\mathrm{i}et \sum_i (c_i^\dagger c_{i+1} - c_{i+1}^\dagger c_i)$. One can also calculate the ω-integrated spectral weight:

$$S^{\mathrm{reg}}(\omega) = \int_{0^+}^{\omega} \mathrm{d}\omega' \sigma^{\mathrm{reg}}(\omega'). \tag{3.93}$$

The optical conductivity, $\sigma^{\mathrm{reg}}(\omega)$, and its integral $S^{\mathrm{reg}}(\omega)$ for the 1D Holstein model are presented in Fig. 3.14 [35]. The upper panel (a) gives the results for intermediate-to-strong EP coupling, i.e., near the polaron crossover, in the adiabatic regime. The main signature of $\sigma^{\mathrm{reg}}(\omega)$ is that the spectrum is strongly asymmetric, which is characteristic for rather large polarons (Sect. 2.4.6). Importantly, the weaker decay at the high-energy side meets the experimental findings for many polaronic materials like $\mathrm{TiO_2}$ [260] even better than standard small-polaron theory.

At larger EPI, $\lambda = 2$ and $\omega_0 = 0.4t$, a more pronounced and symmetric maximum appears in the low-temperature optical response (see Fig. 3.14b). The maximum is located below the analytical one for small polarons at $T = 0$. A deviations from the analytical small-polaron result (dashed line in Fig. 3.14b) might be important for relating theory to experiment. When the phonon frequency becomes comparable to the electron transfer amplitude

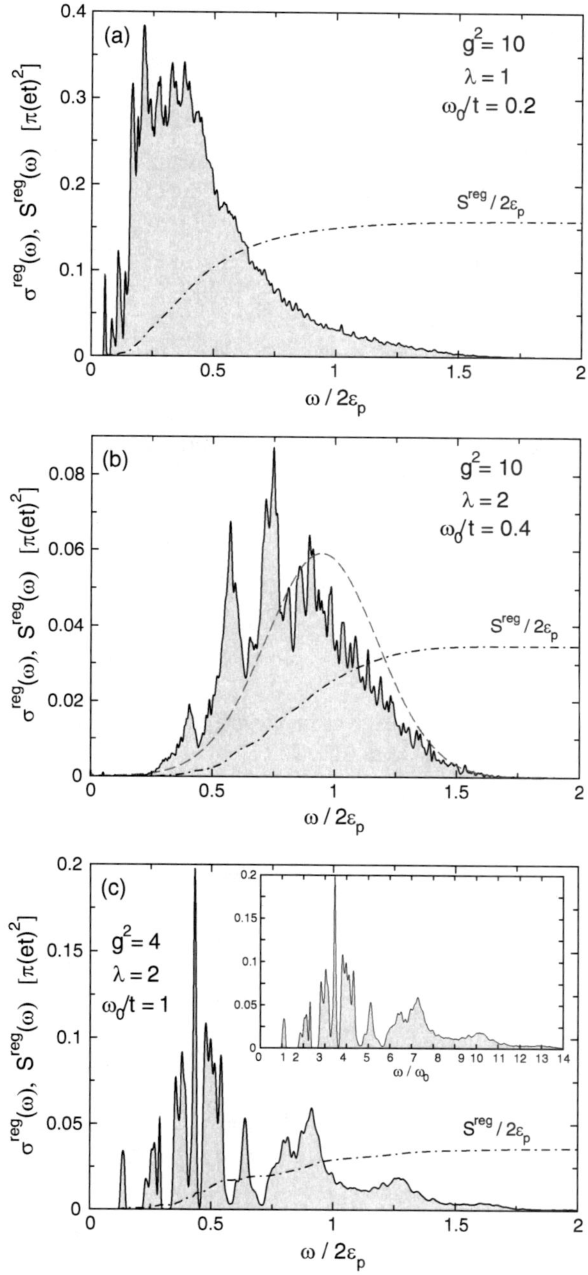

Fig. 3.14. Optical conductivity of the 1D Holstein polaron at $T = 0$ (after [35]). Results for $\sigma^{\text{reg}}(\omega)$ and $S^{\text{reg}}(\omega)$ are obtained by VED. The *dashed line* in (**b**) displays the analytical small-polaron result (cf. [247]). Here $\varepsilon_{\text{p}} \equiv E_{\text{p}}$

(Fig. 3.14c) different absorption bands appear, which can be classified according to the number of phonons involved in the optical transition (see inset).

3.8.3 Spectral and Green's Functions

The polaron problem has the exact solution (3.41) in the extreme infinite-coupling limit, $\lambda = \infty$, for any EPI conserving the on-site occupation numbers of electrons. For a finite but strong coupling $1/\lambda$ perturbation expansion is applied. Importantly, the analytical perturbation theory becomes practically exact in a wider range of the adiabatic parameter and of the coupling constant for the long-range Fröhlich interaction (Sect. 3.4.3).

Keeping this in mind, let us calculate the polaron spectral function in the first order in $1/\lambda$ [261,262]. We can neglect $H_{\text{p-ph}}$ in the first order if $1/\lambda \ll 1$. To understand spectral properties of a single polaron, we also neglect the polaron–polaron interaction. Then, the energy levels are

$$E_{\tilde{m}} = \sum_{\mathbf{k}} \xi_{\mathbf{k}} n_{\mathbf{k}} + \sum_{\mathbf{q}} \omega_{\mathbf{q}} [n_{\mathbf{q}} + 1/2], \tag{3.94}$$

where $\xi_{\mathbf{k}}$ is the small-polaron band dispersion with respect to the chemical potential, and the transformed eigenstates $|\tilde{m}\rangle$ are sorted by the polaron Bloch-state occupation numbers, $n_{\mathbf{k}} = 0, 1$, and the phonon occupation numbers, $n_{\mathbf{q}} = 0, 1, 2, \ldots, \infty$.

The spectral function of any system described by quantum numbers m, n with the eigenvalues E_n, E_m is defined as

$$A(\mathbf{k}, \omega) \equiv \pi (1 + \mathrm{e}^{-\omega/T}) \mathrm{e}^{\Omega/T} \sum_{n,m} \mathrm{e}^{-E_n/T} \left|\langle n| c_{\mathbf{k}} |m\rangle\right|^2 \delta(\omega_{nm} + \omega). \tag{3.95}$$

It is real and positive, $A(\mathbf{k}, \omega) > 0$, and obeys the sum rule:

$$\frac{1}{\pi} \int_{-\infty}^{\infty} \mathrm{d}\omega A(\mathbf{k}, \omega) = 1. \tag{3.96}$$

The matrix elements of the electron operators can be written as

$$\langle n| c_{\mathbf{k}} |m\rangle = \frac{1}{\sqrt{N}} \sum_{\mathbf{m}} \mathrm{e}^{-\mathrm{i}\mathbf{k}\cdot\mathbf{m}} \langle \tilde{n}| c_i \hat{X}_i |\tilde{m}\rangle \tag{3.97}$$

using the Wannier representation and the Lang–Firsov transformation. Now, applying the Fourier transform of the δ-function, the spectral function is expressed as

$$A(\mathbf{k}, \omega) = \frac{1}{2} \int_{-\infty}^{\infty} \mathrm{d}t \mathrm{e}^{\mathrm{i}\omega t} \frac{1}{N} \sum_{\mathbf{m},\mathbf{n}} \mathrm{e}^{\mathrm{i}\mathbf{k}\cdot(\mathbf{n}-\mathbf{m})} \times \left\{ \langle\langle c_i(t) \hat{X}_i(t) c_j^\dagger \hat{X}_j^\dagger \rangle\rangle + \langle\langle c_j^\dagger \hat{X}_j^\dagger c_i(t) \hat{X}_i(t) \rangle\rangle \right\}. \tag{3.98}$$

Here, the quantum and statistical averages are performed for independent polarons and phonons; therefore, $\langle\langle c_i(t)\hat{X}_i(t)\hat{X}_j^\dagger c_i^\dagger\rangle\rangle = \langle\langle c_i(t)c_j^\dagger\rangle\rangle\langle\langle\hat{X}_i(t)\hat{X}_j^\dagger\rangle\rangle$. The Heisenberg free-polaron operator evolves with time as $c_{\mathbf{k}}(t) = c_{\mathbf{k}}\exp(-\mathrm{i}\xi_{\mathbf{k}}t)$, so that

$$\langle\langle c_i(t)c_i^\dagger\rangle\rangle = \frac{1}{N}\sum_{\mathbf{k}',\mathbf{k}''} \mathrm{e}^{\mathrm{i}(\mathbf{k}'\cdot\mathbf{m}-\mathbf{k}''\cdot\mathbf{n})}\langle\langle c_{\mathbf{k}'}(t)c_{\mathbf{k}''}^\dagger\rangle\rangle \tag{3.99}$$

$$= \frac{1}{N}\sum_{\mathbf{k}'}[1-\bar{n}(\mathbf{k}')]\mathrm{e}^{\mathrm{i}\mathbf{k}'\cdot(\mathbf{m}-\mathbf{n})-\mathrm{i}\xi_{\mathbf{k}'}t},$$

$$\langle\langle c_i^\dagger c_i(t)\rangle\rangle = \frac{1}{N}\sum_{\mathbf{k}'}\bar{n}(\mathbf{k}')\mathrm{e}^{\mathrm{i}\mathbf{k}'\cdot(\mathbf{m}-\mathbf{n})-\mathrm{i}\xi_{\mathbf{k}'}t}, \tag{3.100}$$

where $\bar{n}(\mathbf{k}) = [1+\exp\xi_{\mathbf{k}}/T]^{-1}$ is the Fermi–Dirac distribution function of polarons. The Heisenberg free-phonon operator evolves in a similar way, $d_{\mathbf{q}}(t) = d_{\mathbf{q}}\exp(-\mathrm{i}\omega_{\mathbf{q}}t)$, so that

$$\langle\langle\hat{X}_i(t)\hat{X}_j^\dagger\rangle\rangle = \prod_{\mathbf{q}}\langle\langle\exp[u_i(\mathbf{q},t)d_{\mathbf{q}} - \mathrm{H.c.}]\exp[-u_j(\mathbf{q})d_{\mathbf{q}} - \mathrm{H.c.}]\rangle\rangle, \tag{3.101}$$

where $u_{i,j}(\mathbf{q},t) = u_{i,j}(\mathbf{q})\exp(-\mathrm{i}\omega_{\mathbf{q}}t)$. This average is calculated using the operator identity, as in (3.46):

$$\langle\langle\hat{X}_i(t)\hat{X}_j^\dagger\rangle\rangle = \exp\left\{-\frac{1}{2N}\sum_{\mathbf{q}}|\gamma(\mathbf{q})|^2 f_{\mathbf{q}}(\mathbf{m}-\mathbf{n},t)\right\}, \tag{3.102}$$

where

$$f_{\mathbf{q}}(\mathbf{m},t) = [1-\cos(\mathbf{q}\cdot\mathbf{m})\cos(\omega_{\mathbf{q}}t)]\coth\frac{\omega_{\mathbf{q}}}{2T} + \mathrm{i}\cos(\mathbf{q}\cdot\mathbf{m})\sin(\omega_{\mathbf{q}}t). \tag{3.103}$$

Here, we used the symmetry of $\gamma(-\mathbf{q}) = \gamma(\mathbf{q})$, so that terms containing $\sin(\mathbf{q}\cdot\mathbf{m})$ have disappeared.

To proceed with the analytical results, we consider low temperatures, $T \ll \omega_{\mathbf{q}}$, when $\coth(\omega_{\mathbf{q}}/2T) \approx 1$. Expanding the exponent in (3.102) and performing summation over $\mathbf{m}, \mathbf{n}$, $\mathbf{k}'$ and integration over time, we arrive at [262]

$$A(\mathbf{k},\omega) = \sum_{l=0}^{\infty}\left[A_l^{(-)}(\mathbf{k},\omega) + A_l^{(+)}(\mathbf{k},\omega)\right], \tag{3.104}$$

where

$$A_l^{(-)}(\mathbf{k},\omega) = \pi Z\sum_{\mathbf{q}_1,\ldots,\mathbf{q}_l}\frac{\prod_{r=1}^{l}|\gamma(\mathbf{q}_r)|^2}{(2N)^l l!} \tag{3.105}$$

$$\times\left[1-\bar{n}\left(\mathbf{k}-\sum_{r=1}^{l}\mathbf{q}_r\right)\right]\delta\left(\omega-\sum_{r=1}^{l}\omega_{\mathbf{q}_r}-\xi_{\mathbf{k}-\sum_{r=1}^{l}\mathbf{q}_r}\right),$$

$$A_l^{(+)}(\mathbf{k},\omega) = \pi Z \sum_{\mathbf{q}_1,\ldots,\mathbf{q}_l} \frac{\prod_{r=1}^{l} |\gamma(\mathbf{q}_r)|^2}{(2N)^l l!} \times \bar{n}\left(\mathbf{k}+\sum_{r=1}^{l}\mathbf{q}_r\right) \delta\left(\omega + \sum_{r=1}^{l}\omega_{\mathbf{q}_r} - \xi_{\mathbf{k}+\sum_{r=1}^{l}\mathbf{q}_r}\right), \tag{3.106}$$

and $Z = \exp\left(-(2N)^{-1}\sum_{\mathbf{q}} |\gamma(\mathbf{q})|^2\right)$.

Clearly (3.104) is in the form of the perturbative multiphonon expansion. Each contribution $A_l^{(\pm)}(\mathbf{k},\omega)$ to the spectral function describes the transition from the initial state $\mathbf{k}$ of the polaron band to the final state $\mathbf{k} \pm \sum_{r=1}^{l} \mathbf{q}_r$ with the emission (or absorption) of l phonons.

The $1/\lambda$ expansion result (3.104) is different from the conventional spectral function of metallic electrons coupled to phonons in the Migdal–Eliashberg theory [263,264]. There is no imaginary part of the self-energy since the exponentially small (at low temperatures) polaronic damping (3.56) is neglected. Instead EPI leads to the coherent dressing of electrons by phonons, and phonon "sidebands." The spectral function of the polaronic carriers comprises two different parts. The first ($l = 0$) $\mathbf{k}$-dependent *coherent* term arises from the polaron band tunneling:

$$A_{\mathrm{coh}}(\mathbf{k},\omega) = \left[A_0^{(-)}(\mathbf{k},\omega) + A_0^{(+)}(\mathbf{k},\omega)\right] = \pi Z \delta(\omega - \xi_{\mathbf{k}}). \tag{3.107}$$

The spectral weight of the coherent part is suppressed as $Z \ll 1$. However in the case of the Fröhlich interaction, the effective mass is less enhanced, since the band-narrowing factor Z' (3.77) in $\xi_{\mathbf{k}} = Z'E_{\mathbf{k}} - \mu$ is large compared with Z [262].

The second *incoherent* part $A_{\mathrm{incoh}}(\mathbf{k},\omega)$ comprises all the terms with $l \geq 1$. It describes excitations accompanied by emission and absorption of phonons. We notice that its spectral density spreads over a wide energy range of about twice the polaron level shift E_{p}, which might be larger than the unrenormalized bandwidth 2D in the rigid lattice without phonons. On the contrary, the coherent part shows a dispersion only in the energy window of the order of the polaron bandwidth, $w = Z'D$. It is interesting that there is some $\mathbf{k}$ dependence of the *incoherent* background as well, if the EPI matrix element and/or phonon frequencies depend on $\mathbf{q}$. Only in the Holstein model with the short-range dispersionless e–ph interaction $\gamma(\mathbf{q}) = \gamma_0$ and $\omega_{\mathbf{q}} = \omega_0$, the incoherent part is momentum independent (see also [7]):

$$A_{\mathrm{incoh}}(\mathbf{k},\omega) = \pi \frac{Z}{N} \sum_{l=1}^{\infty} \frac{\gamma_0^{2l}}{2^l l!} \times \sum_{\mathbf{k}'} \left\{\left[1 - \bar{n}(\mathbf{k}')\right] \delta(\omega - l\omega_0 - \xi_{\mathbf{k}'}) + \bar{n}(\mathbf{k}')\, \delta(\omega + l\omega_0 - \xi_{\mathbf{k}'})\right\}. \tag{3.108}$$

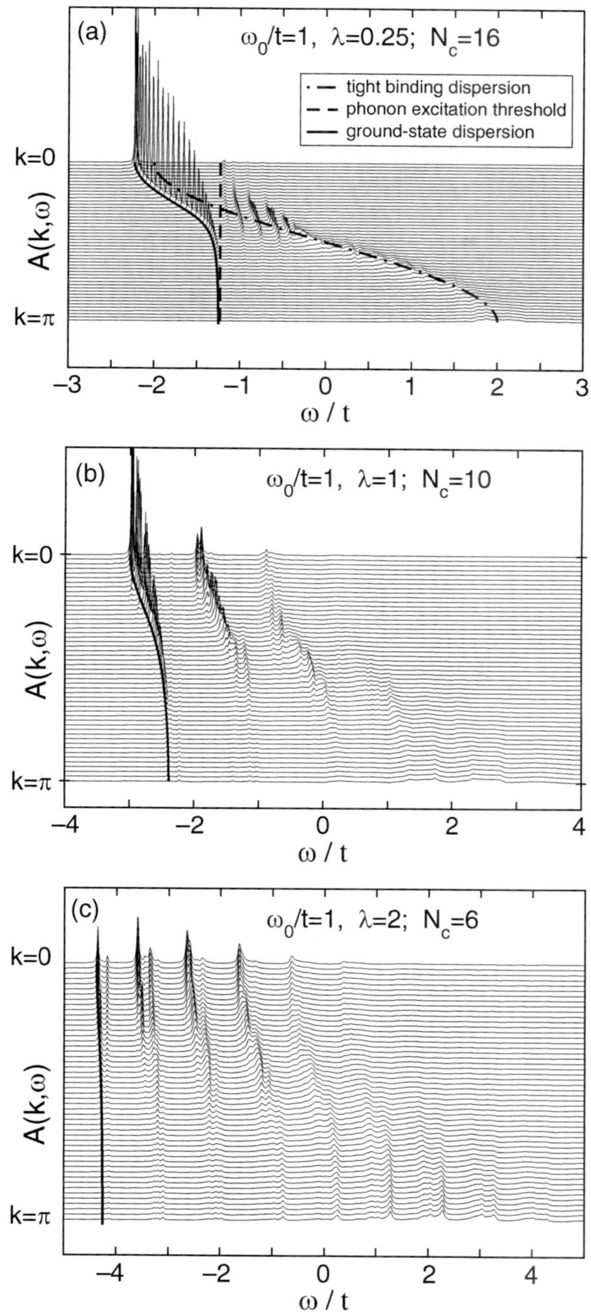

Fig. 3.15. Spectral function of the 1D Holstein polaron in the weak (**a**), intermediate (**b**), and strong (**c**) EP coupling regimes based on finite-cluster ED with N_c sites, and $M = 7$ ($\lambda = 0.25$), $M = 15$ ($\lambda = 1$), $M = 25$ ($\lambda = 2$) phonon quanta (after [35])

As soon as we know the spectral function, different Green's functions (GF) are readily obtained using their analytical properties. For example, the temperature GF is given by the integral

$$\mathcal{G}(\mathbf{k},\omega_k) = \frac{1}{\pi}\int_{-\infty}^{\infty} \mathrm{d}\omega' \frac{A(\mathbf{k},\omega')}{\mathrm{i}\omega_k - \omega'}, \tag{3.109}$$

where $\omega_k = \pi T(2k+1)$, $k = 0, \pm 1, \pm 2, \ldots$. Calculating the integral, we find in the Holstein model [261]

$$\mathcal{G}(\mathbf{k},\omega_n) = \frac{Z}{\mathrm{i}\omega_n - \xi_{\mathbf{k}}} + \frac{Z}{N}\sum_{l=1}^{\infty}\frac{\gamma_0^{2l}}{2^l l!}\sum_{\mathbf{k}'}\left\{\frac{1-\bar{n}\left(\mathbf{k}'\right)}{\mathrm{i}\omega_n - l\omega_0 - \xi_{\mathbf{k}'}} + \frac{\bar{n}\left(\mathbf{k}'\right)}{\mathrm{i}\omega_n + l\omega_0 - \xi_{\mathbf{k}'}}\right\}. \tag{3.110}$$

Here, the first term describes the coherent tunneling in the narrow polaron band while the second **k**-independent sum describes the phonon sidebands.

The spectral function (3.110) satisfies the major sum rule (3.96). However, the higher-momentum integrals, $\int_{-\infty}^{\infty} \mathrm{d}\omega\omega^p A(\mathbf{k},\omega)$ with $p > 0$, calculated using (3.110), differ from the exact values [265] by an amount proportional to $1/\lambda$. The difference is due to a partial "undressing" of high-energy excitations in the sidebands, which is beyond the first-order $1/\lambda$ expansion. A rather accurate Green's function of the Holstein polaron has been recently obtained by summing all the diagrams, but with each diagram averaged over its free propagators' momenta [266, 267]. The resulting Green's function satisfies exactly the first six spectral weight sum rules. All higher sum rules are satisfied with great accuracy, becoming asymptotically exact for coupling both much larger and much smaller than the free particle bandwidth.

The spectral properties of the Holstein model in a wide parameter range have been studied numerically using ED (see [35, 156, 161, 189, 226, 256, 257, 259, 268], and references therein). Figure 3.15a shows that at the weak EPI, the electronic spectrum of 1D Holstein model is nearly unaffected for energies below the phonon emission threshold. Hence, for the case considered with ω_0 lying inside the bare-electron band $E_k = -2t\cos ka$, the signal corresponding to the renormalized dispersion ϵ_k nearly coincides with the tight-binding cosine band (shifted $\propto E_\mathrm{p}$) up to some k_X, where the phonon intersects the bare-electron band with the familiar flattening effect (Sect. 3.5). Reaching the intermediate EPI (polaron crossover) regime, a coherent band separates from the rest of the spectrum ($k_X \to \pi$; see Fig. 3.15b). At the same time, its spectral weight becomes smaller and will be transferred to the incoherent part, where several subbands emerge.

$A(\mathbf{k},\omega)$ in the strong-coupling case is shown in Fig. 3.15c. The coherent quasiparticle absorption band becomes extremely narrow and its bandwidth approaches the strong-coupling result. The incoherent part of the spectrum carries most of the spectral weight consisting of a sequence of subbands separated in energy by ω_0, in agreement with the analytical results.

Effects of the finite-range EPI on the spectral properties of lattice polarons have been studied numerically by Fehske *et al.* [269] using exact Lanczos diagonalization method in the framework of the 1D model of [60]. The polaron band structure has been calculated in agreement with the analytical and CTQMC results (Sect. 3.4.3). The optical absorption of lattice polarons with a finite-range (Fröhlich-type) EPI has been found similar to the continuous-polaron absorption (Sect. 3.4.3) for all EPI strengths.

The polaron features due to EPIs with different coupling ranges were also investigated in the framework of the variational approach [270]. The ground-state energy, the spectral weight, the average kinetic energy, the mean number of phonons, and the electron–lattice correlation function were calculated for the system with coupling to local and nearest-neighbor lattice displacements and compared with the long-range case. As in [262], a substantially different mass renormalization compared with the coherent weight reduction, $Z \ll Z'$, was found for the finite-range EPI.

4

Bipolaron

4.1 Polaron–Polaron Interaction

Polarons interact with each other, cf. (3.40) for small polarons. The range of the deformation surrounding the Fröhlich polarons is quite large, and their deformation fields are overlapped at finite density. Taking into account both the long-range attraction of polarons owing to the lattice deformations *and* their direct Coulomb repulsion, the residual *long-range* interaction turns out rather weak and repulsive in ionic crystals [26]. In the long-wavelength limit ($q \ll \pi/a$), the Fröhlich EPI dominates in the attractive part, but polarons repel each other at large distances, $|\mathbf{m}-\mathbf{n}| \gg a$:

$$v(\mathbf{m}-\mathbf{n}) = \frac{e^2}{\varepsilon_0|\mathbf{m}-\mathbf{n}|} > 0. \tag{4.1}$$

The Fröhlich EPI nearly nullifies the bare Coulomb repulsion, if $\varepsilon_0 \gg 1$, but cannot overscreen it at large distances.

Considering the polaron–phonon interaction in the multipolaron system, we have to take into account the dynamic properties of the polaron response function [136]. One may erroneously believe that the long-range Fröhlich EPI becomes a short-range (Holstein) EPI due to screening of ionic potentials by heavy polaronic carriers. In fact, small polarons cannot screen high-frequency optical vibrations because their renormalized plasma frequency is comparable with or even less than the phonon frequency in the strong-coupling and/or the dilute (low-density) regimes. In the absence of bipolarons (see below), one can apply the ordinary bubble approximation to calculate the dielectric response function of polarons at the frequency ω as [136]

$$\epsilon(\mathbf{q},\omega) = 1 - 2v(\mathbf{q})\sum_{\mathbf{k}} \frac{\bar{n}(\mathbf{k}+\mathbf{q}) - \bar{n}(\mathbf{k})}{\omega - \epsilon_{\mathbf{k}} + \epsilon_{\mathbf{k}+\mathbf{q}}}. \tag{4.2}$$

A.S. Alexandrov and J.T. Devreese, *Advances in Polaron Physics*, Springer Series in Solid-State Sciences 159,
DOI: 10.1007/978-3-642-01896-1_4,

This expression describes the response of polarons to the external field of frequency $\omega \leq \omega_0$, when phonons in the polaron cloud follow the polaron motion. In the static limit, we obtain the usual Debye screening at large distances ($q \to 0$). For the temperature larger than the polaron half-bandwidth, $T > w$, one can approximate the polaron distribution function as

$$\bar{n}(\mathbf{k}) \approx \frac{n_{\mathrm{p}}}{2a^3}\left(1 - \frac{(2-n_{\mathrm{p}})\epsilon_{\mathbf{k}}}{2T}\right), \tag{4.3}$$

to obtain

$$\epsilon(q,0) = 1 + \frac{q_{\mathrm{s}}^2}{q^2}, \tag{4.4}$$

where

$$q_{\mathrm{s}} = \left[\frac{2\pi e^2 n_{\mathrm{p}}(2-n_{\mathrm{p}})}{\epsilon_0 T a^3}\right]^{1/2},$$

and n_{p} is the number of polarons per unit cell. However, for a finite but rather low frequency, $\omega_0 > \omega \gg w$, the polaron response becomes dynamic:

$$\epsilon(\mathbf{q},\omega) = 1 - \frac{\omega_{\mathrm{p}}^2(\mathbf{q})}{\omega^2}, \tag{4.5}$$

where

$$\omega_{\mathrm{p}}^2(\mathbf{q}) = 2v(\mathbf{q})\sum_{\mathbf{k}} \bar{n}(\mathbf{k})(\epsilon_{\mathbf{k}+\mathbf{q}} - \epsilon_{\mathbf{k}}) \tag{4.6}$$

is the temperature-dependent *polaron* plasma frequency squared. The polaron plasma frequency is rather low due to the large static dielectric constant, $\epsilon_0 \gg 1$, and the enhanced polaron mass $m^* \gg m_{\mathrm{e}}$. If $\omega_0 > \omega_{\mathrm{p}}$, the singular behavior of the Fröhlich EPI, $\gamma(\mathbf{q}) \sim 1/q$, is unaffected by screening. Polarons are too slow to screen high-frequency crystal field oscillations. As a result, EPI with high-frequency optical phonons in ionic solids remains unscreened at any density of polarons.

Another important issue is a possibility of the Wigner crystallization of polarons. Because the net long-range interaction is relatively weak, a relevant dimensionless parameter $r_{\mathrm{s}} = m^* a e^2/\epsilon_0(4\pi n_{\mathrm{p}}/3)^{1/3}$ is not very large in ionic semiconductors. The Wigner crystallization appears around $r_{\mathrm{s}} \simeq 100$ or larger, which corresponds to the atomic density of polarons $n_{\mathrm{p}} \leq 10^{-6}$ with $\epsilon_0 = 30$ and $m^* = 5m_{\mathrm{e}}$. This estimate tells us that polaronic carriers are in a liquid state [136] at substantial doping levels, but can be crystallized at low doping (see Sect. 5.1).

When the short-range deformation and molecular-type (i.e., Holstein) EPIs are added to the Fröhlich interaction, two polarons attract each other at a short distance of about the lattice constant. Then, owing to a narrow band, two lattice polarons easily form a local tightly bound state, i.e., a *small* bipolaron [271–276]. One can estimate the coupling constant λ and the adiabatic ratio ω_0/t, at which the small bipolaron "instability" occurs [136]. The characteristic attractive potential is $|v| = D(\lambda - \mu_c)$, where μ_c is the dimensionless Coulomb repulsion and λ includes the interaction with all phonon modes. The radius of the potential is about a. In three dimensions, a bound state of two attractive particles appears, if $|v| \geq \pi^2/8m^*a^2$. Substituting the polaron mass, $m^* = [2a^2t]^{-1}\exp(\gamma\lambda D/\omega_0)$, we find

$$\frac{t}{\omega_0} \leq (\gamma z\lambda)^{-1} \ln\left[\frac{\pi^2}{4z(\lambda - \mu_c)}\right]. \tag{4.7}$$

As a result, small bipolarons form at $\lambda \geq \mu_c + \pi^2/4z$, which is almost independent of the adiabatic ratio.

4.2 Holstein Bipolaron

The attractive energy of two small polarons is generally much larger than the polaron bandwidth, which allows for a consistent treatment of small bipolarons [130,271,272]. Under this condition, the hopping term in the transformed Hamiltonian $\tilde{H}$ is a small perturbation of the ground state of immobile bipolarons and free phonons:

$$\tilde{H} = H_0 + H_{\text{pert}}, \tag{4.8}$$

where

$$H_0 = \frac{1}{2}\sum_{i,j} v_{ij} c_i^\dagger c_j^\dagger c_j c_i + \sum_{\mathbf{q},v} \omega_{\mathbf{q}v}[d_{\mathbf{q}v}^\dagger d_{\mathbf{q}v} + 1/2] \tag{4.9}$$

and

$$H_{\text{pert}} = \sum_{i,j} \hat{\sigma}_{ij} c_i^\dagger c_j. \tag{4.10}$$

Let us first discuss the dynamics of *on-site* bipolarons, which are the ground state of the system with the Holstein nondispersive EPI [271,273–277]. The on-site bipolaron is formed if $2E_p > U$, where U is the on-site Coulomb correlation energy (the Hubbard U). The intersite polaron–polaron interaction (3.40) is just the Coulomb repulsion since the phonon-mediated attraction between two polarons on different sites is zero in the Holstein model. Two or more on-site bipolarons as well as three or more polarons cannot occupy the same site because of the Pauli exclusion principle. Hence, bipolarons repel single polarons and each other. Their binding energy, $\Delta = 2E_p - U$, is larger than the polaron half-bandwidth, $\Delta \gg w$, so that there are no unbound polarons in the ground state. H_{pert} (4.10) destroys bipolarons in the first order. Hence, it has no diagonal matrix elements. Then, the bipolaron dynamics, including superconductivity, is described by the use of another canonical transformation $\exp(S_2)$ [271], which eliminates the first order of H_{pert}:

$$(S_2)_{fp} = \sum_{i,j} \frac{\langle f|\hat{\sigma}_{ij} c_i^\dagger c_j |p\rangle}{E_f - E_p}. \tag{4.11}$$

Here, $E_{f,p}$ and $|f\rangle, |p\rangle$ are the energy levels and the eigenstates of H_0. Neglecting the terms of the order higher than $(w/\Delta)^2$, one obtains

$$\begin{aligned}(H_\mathrm{b})_{ff'} &\equiv \left(\mathrm{e}^{S_2} \tilde{H}\, \mathrm{e}^{-S_2}\right)_{ff'}, \\ (H_\mathrm{b})_{ff'} &\approx (H_0)_{ff'} - \frac{1}{2} \sum_\nu \sum_{i\neq i', j\neq j'} \langle f|\hat{\sigma}_{ii'} c_i^\dagger c_{i'} |p\rangle \langle p|\hat{\sigma}_{jj'} c_j^\dagger c_{j'} |f'\rangle \\ &\times \left(\frac{1}{E_\mathrm{p} - E_{f'}} + \frac{1}{E_\mathrm{p} - E_f} \right).\end{aligned} \tag{4.12}$$

The *bipolaronic* Hamiltonian H_b is defined in the subspace $|f\rangle, |f'\rangle$ with no single (unbound) polarons. On the other hand, the intermediate bra $\langle p|$ and ket $|p\rangle$ refer to configurations involving two unpaired polarons and any number of phonons. Hence, we have

$$E_\mathrm{p} - E_f = \Delta + \sum_{\mathbf{q},v} \omega_{\mathbf{q}v} \left(n_{\mathbf{q}v}^\mathrm{p} - n_{\mathbf{q}v}^f\right), \tag{4.13}$$

where $n_{\mathbf{q}v}^{f,\mathrm{p}}$ are the phonon occupation numbers $(0, 1, 2, 3, \ldots, \infty)$.

The lowest eigenstates of H_b are in the subspace, which has only doubly occupied $c_{\mathbf{m}s}^\dagger c_{\mathbf{m}s'}^\dagger |0\rangle$ or empty $|0\rangle$ sites. On-site bipolaron tunneling is a two-step transition. It takes place via a single polaron tunneling to a neighboring site. The subsequent tunneling of its "partner" to the same site restores the initial energy state of the system. There are no *real* phonons emitted or absorbed because the (bi)polaron band is narrow. Hence, we can average H_b with respect to phonons:

$$\begin{aligned}H_\mathrm{b} = H_0 - \mathrm{i} &\sum_{\mathbf{m}\neq\mathbf{m}', s} \sum_{\mathbf{n}\neq\mathbf{n}', s'} t(\mathbf{m} - \mathbf{m}') t(\mathbf{n} - \mathbf{n}') \\ &\times c_{\mathbf{m}s}^\dagger c_{\mathbf{m}'s} c_{\mathbf{n}s'}^\dagger c_{\mathbf{n}'s'} \int_0^\infty \mathrm{d}t \mathrm{e}^{-\mathrm{i}\Delta t} \Phi_{\mathbf{mm}'}^{\mathbf{nn}'}(t).\end{aligned} \tag{4.14}$$

Here, $\Phi_{\mathbf{mm}'}^{\mathbf{nn}'}(t)$ is a multiphonon correlator:

$$\Phi_{\mathbf{mm}'}^{\mathbf{nn}'}(t) \equiv \langle\langle \hat{X}_i^\dagger(t) \hat{X}_{i'}(t) \hat{X}_j^\dagger \hat{X}_{j'} \rangle\rangle, \tag{4.15}$$

which can be readily calculated as [131]

$$\begin{aligned}\Phi_{\mathbf{mm}'}^{\mathbf{nn}'}(t) &= \mathrm{e}^{-g^2(\mathbf{m}-\mathbf{m}')} \mathrm{e}^{-g^2(\mathbf{n}-\mathbf{n}')} \\ &\times \exp\left\{ \frac{1}{2N} \sum_{\mathbf{q},v} |\gamma(\mathbf{q}, v)|^2 F_\mathbf{q}(\mathbf{m}, \mathbf{m}', \mathbf{n}, \mathbf{n}') \frac{\cosh\left[\omega_{\mathbf{q}v}\left(\frac{1}{2T} - \mathrm{i}t\right)\right]}{\sinh\left[\frac{\omega_{\mathbf{q}\nu}}{2T}\right]} \right\},\end{aligned} \tag{4.16}$$

where $F_{\mathbf{q}}(\mathbf{m},\mathbf{m}',\mathbf{n},\mathbf{n}') = \cos[\mathbf{q}\cdot(\mathbf{n}'-\mathbf{m})] + \cos[\mathbf{q}\cdot(\mathbf{n}-\mathbf{m}')] - \cos[\mathbf{q}\cdot(\mathbf{n}'-\mathbf{m}')] - \cos[\mathbf{q}\cdot(\mathbf{n}-\mathbf{m})]$.

Taking into account that there are only bipolarons in the subspace, where H_{b} operates, one can rewrite the Hamiltonian in terms of the creation $b_{\mathbf{m}}^{\dagger} = c_{\mathbf{m}\uparrow}^{\dagger}c_{\mathbf{m}\downarrow}^{\dagger}$ and annihilation $b_{\mathbf{m}} = c_{\mathbf{m}\downarrow}c_{\mathbf{m}\uparrow}$ operators of singlet pairs [271]:

$$\begin{aligned} H_{\mathrm{b}} = &-\sum_{\mathbf{m}}\left[\Delta + \frac{1}{2}\sum_{\mathbf{m}'} v^{(2)}(\mathbf{m}-\mathbf{m}')\right] n_{\mathbf{m}} \\ &+ \sum_{\mathbf{m}\neq\mathbf{m}'}\left[t_{\mathrm{b}}(\mathbf{m}-\mathbf{m}')b_{\mathbf{m}}^{\dagger}b_{\mathbf{m}'} + \frac{1}{2}\bar{v}(\mathbf{m}-\mathbf{m}')n_{\mathbf{m}}n_{\mathbf{m}'}\right]. \end{aligned} \tag{4.17}$$

Here, $n_{\mathbf{m}} = b_{\mathbf{m}}^{\dagger}b_{\mathbf{m}}$ is the bipolaron site occupation operator, $\bar{v}(\mathbf{m}-\mathbf{m}') = 4v(\mathbf{m}-\mathbf{m}') + v^{(2)}(\mathbf{m}-\mathbf{m}')$ is the bipolaron–bipolaron interaction including the direct polaron–polaron interaction $v(\mathbf{m}-\mathbf{m}')$ and a repulsive correlation energy:

$$v^{(2)}(\mathbf{m}-\mathbf{m}') = 2\mathrm{i}\int_0^{\infty} \mathrm{d}\tau \mathrm{e}^{-\mathrm{i}\Delta\tau}\Phi_{\mathbf{m}\mathbf{m}'}^{\mathbf{m}'\mathbf{m}}(\tau). \tag{4.18}$$

This additional repulsion appears since a virtual hop of one of two polarons of the pair is forbidden, if the neighboring site is occupied by another pair. The bipolaron transfer integral, t_{b} is of the second order in $t(\mathbf{m})$:

$$t_{\mathrm{b}}(\mathbf{m}-\mathbf{m}') = -2\mathrm{i}t^2(\mathbf{m}-\mathbf{m}')\int_0^{\infty} \mathrm{d}\tau \mathrm{e}^{-\mathrm{i}\Delta\tau}\Phi_{\mathbf{m}\mathbf{m}'}^{\mathbf{m}\mathbf{m}'}(\tau). \tag{4.19}$$

The multiphonon correlator is simplified for dispersionless phonons at $T \ll \omega_0$ as

$$\begin{aligned} \Phi_{\mathbf{m}\mathbf{m}'}^{\mathbf{m}\mathbf{m}'}(t) &= \mathrm{e}^{-2g^2(\mathbf{m}-\mathbf{m}')}\exp\left[-2g^2(\mathbf{m}-\mathbf{m}')\mathrm{e}^{-\mathrm{i}\omega_0 t}\right], \\ \Phi_{\mathbf{m}\mathbf{m}'}^{\mathbf{m}'\mathbf{m}}(t) &= \mathrm{e}^{-2g^2(\mathbf{m}-\mathbf{m}')}\exp\left[2g^2(\mathbf{m}-\mathbf{m}')\mathrm{e}^{-\mathrm{i}\omega_0 t}\right], \end{aligned}$$

which yields [278]

$$t(\mathbf{m}) = -\frac{2t^2(\mathbf{m})}{\Delta}\mathrm{e}^{-2g^2(\mathbf{m})}\sum_{l=0}^{\infty}\frac{[-2g^2(\mathbf{m})]^l}{l!(1+l\omega_0/\Delta)} \tag{4.20}$$

and

$$v^{(2)}(\mathbf{m}) = \frac{2t^2(\mathbf{m})}{\Delta}\mathrm{e}^{-2g^2(\mathbf{m})}\sum_{l=0}^{\infty}\frac{[2g^2(\mathbf{m})]^l}{l!(1+l\omega_0/\Delta)}. \tag{4.21}$$

When $\Delta \ll \omega_0$, we can keep the first term only with $l = 0$ in the bipolaron hopping integral (4.19). In this case, the bipolaron half-bandwidth $zt(\mathbf{a})$ is of the order of $2w^2/(z\Delta)$. However, if the bipolaron binding energy is large, $\Delta \gg \omega_0$, the bipolaron bandwidth dramatically decreases proportionally to e^{-4g^2} in the limit $\Delta \to \infty$. This limit is not realistic since $\Delta = 2E_{\mathrm{p}} - V_{\mathrm{c}} < 2g^2\omega_0$. In a more realistic regime, $\omega_0 < \Delta < 2g^2\omega_0$ (4.19) yields [278]

$$t_{\mathrm{b}}(\mathbf{m}) \approx \frac{2\sqrt{2\pi}t^2(\mathbf{m})}{\sqrt{\omega_0\Delta}} \exp\left[-2g^2 - \frac{\Delta}{\omega_0}\left(1 + \ln\frac{2g^2(\mathbf{m})\omega_0}{\Delta}\right)\right]. \tag{4.22}$$

On the contrary, the bipolaron–bipolaron repulsion (4.21) has no small exponent in the limit $\Delta \to \infty$, $v^{(2)} \propto D^2/\Delta$. Together with the direct Coulomb repulsion, the second order $v^{(2)}$ ensures stability of bipolarons against clustering.

Interestingly, the high-temperature behavior of the bipolaron bandwidth is just opposite to that of the small-polaron bandwidth. While the polaron band collapses with increasing temperature (3.2.2), the bipolaron band becomes wider [279]

$$t_{\mathrm{b}}(\mathbf{m}) \propto \frac{1}{\sqrt{T}} \exp\left[-\frac{E_{\mathrm{p}} + \Delta}{2T}\right] \tag{4.23}$$

for $T > \omega_0$.

The hopping conductivity of strong-coupling on-site bipolarons in the Holstein–Hubbard model (HHM) was found small compared with the hopping conductivity of thermally excited single polarons [279]. However, as the frequency of the electric field increases, the dominant role in the optical absorption is gradually transferred to bipolarons at low temperatures. Like in the single-polaron case, the bipolaron optical absorption can be estimated using the Franck–Condon principle which states that optical transitions take place without any change in the nuclear configuration. The corresponding analysis [280] shows that the absorption coefficient of light by the on-site bipolaron has three Gaussian peaks located at frequencies $\Omega = 4E_a$, $8E_a - U$, and $16E_a$. The lowest peak corresponds to the absorption by single thermally excited polarons. The highest peak is due to the shakeoff of phonons without dissociation of the bipolaron while the main central peak is the absorption involving dissociation. A generalization of the Franck–Condon principle for the optical absorption of intersite bipolarons with a finite-range EPI was given by Alexandrov and Bratkovsky [281].

The optical absorption and single-particle spectral functions of the bipolaron in 1D HHM in the whole range of parameters were calculated using ED on a two-site cluster with two electrons [256], and more recently in [282] using the cluster perturbation theory (CPT). The latter allowed one to calculate the spectrum at continuous wave vectors and to find pronounced deviations (e.g., noncosine dispersions) of the bipolaron band structure from a simple tight-binding band due to an important contribution from the next-nearest-neighbor hoppings.

Treating phonons classically in the extreme adiabatic limit [274, 277, 283–285] found along with the on-site bipolaron (S_0) also an anisotropic pair of polarons lying on two neighboring sites, i.e., the *intersite* bipolaron (S_1). Such bipolarons were originally hypothesized in [286] to explain the anomalous nuclear magnetic relaxation (NMR) in cuprate superconductors. The intersite bipolaron could take a form of a "quadrisinglet" (QS) in 2D HHM, where the electron density at the central site is 1 and "1/4" on the four nearest-neighboring sites. In a certain region of U, where QS is the ground state, the double-well potential barrier, which usually pins polarons and bipolarons to the lattice, depresses to almost zero, so that adiabatic lattice bipolarons can be rather mobile (see [283] for more details).

Sil [287] investigated the stability of the bipolaronic phase in HHM using a modified Lang–Firsov variational transformation with on-site and nearest-neighbor lattice distortions. There is a critical on-site Hubbard repulsion U_c below which the bipolaronic phases are stable for a fixed electron–phonon coupling. In the absence of on-site repulsion, bipolaronic phases are stable over the entire range of electron–phonon coupling for one dimension, whereas there is a critical electron–phonon coupling for formation of a stable bipolaron in two and three dimensions. Mobile S_1 bipolarons were found in 1D HHM using variational methods also in the non- and near-adiabatic regimes with dynamical quantum phonons [288, 289]. The intersite bipolaron with a relatively small effective mass is stable in a wide region of the parameters of HHM due to both exchange and nonadiabaticity effects [289]. Near the strong-coupling limit, the mobile S_1 bipolaron has an effective mass of the order of a single Holstein polaron mass, so that one should not rule out the possibility of a superconducting state of S_1 bipolarons with s- or d-wave symmetry in HHM [288]. The recent diagrammatic Monte Carlo study [183] found S_1 bipolarons for large U at intermediate and large EPI and established the phase diagram of 2D HHM, comprising large and small unbound polarons, S_0 and S_1 domains. Macridin *et al.* [183] emphasized that the transition to the bound state and the properties of the bipolaron in HHM are very different from bound states in the attractive (negative U) Hubbard model without EPI [290].

The two-dimensional many-body HHM was examined within a fluctuation-based effective cumulant approach by Hakioglu and Zhuravlev [291], confirming that the numerical results on systems with finite degrees of freedom (Sect. 3.4) can be qualitatively extended to the systems with large degrees of freedom. When the electron–electron repulsion U is dominant, the transition is to a Mott insulator; when EPI dominates, the transition is to a bipolaronic state. In the former case, the transition was found to be second order in contrast to the transition to the bipolaronic state, which is first order for larger values of U [292].

For a very strong electron–phonon coupling, polarons become self-trapped on a single lattice site and bipolarons are on-site singlets. A finite bipolaron mass appears only in the second order of polaron hopping (4.19), so that on-site bipolarons might be very heavy in the Holstein model with the short-range EPI. As a result, the model led some authors to the conclusion that the formation of itinerant small polarons and bipolarons in real materials is unlikely [139, 185], and high-temperature bipolaronic superconductivity is impossible [293, 294].

4.3 Continuum Fröhlich Bipolaron

While polarons repel each other at large distances, two *large* polarons can also be bound into a *large* bipolaron by an exchange interaction even without any additional EPI but the Fröhlich one [295–304] (see also the reviews [305,306]). Large bipolarons in the continuum limit are referred to as *Fröhlich bipolarons.* Besides the Fröhlich coupling constant, α, the Fröhlich bipolaron energy depends also on the dimensionless parameter U, a measure for the strength of the Coulomb repulsion between the two electrons [301, 302], $U = (e^2/\varepsilon\omega_0)\sqrt{m\omega_0}$. In the discussion of bipolarons, the ratio $\eta = \varepsilon/\varepsilon_0$ of the high-frequency (electronic) and static dielectric constants is important ($0 \leq \eta \leq 1$). The following relation exists between U and α: $U = \sqrt{2}\alpha/(1-\eta)$. Only values of U satisfying the inequality $U \geq \sqrt{2}\alpha$ have a physical meaning. It was shown that bipolaron formation is favored by larger values of α and by smaller values of η.

Intuitive arguments suggesting that the Fröhlich bipolaron is stabilized in going from 3D to 2D had been given, but the first quantitative analysis based on the Feynman path integral was presented in [301,302]. The conventional condition for bipolaron stability is $E_{\mathrm{bip}} \leq 2E_{\mathrm{pol}}$, where E_{pol} and E_{bip} denote the ground-state energies of the polaron and bipolaron at rest, respectively. From this condition, it follows that the Fröhlich bipolaron with zero spin is stable (given the effective Coulomb repulsion between electrons) if the electron–phonon coupling constant is larger than a certain critical value: $\alpha \geq \alpha_{\mathrm{c}}$.

A "phase diagram" for the two continuum polarons – bipolaron system – is shown in Fig. 4.1 for 3D and for 2D. The Fröhlich coupling constant as high as 6.8 is needed to allow for bipolaron formation in 3D. The confinement of the bipolaron in two dimensions facilitates bipolaron formation at smaller α. This can be shown using the PD-type scaling relation between the free energies F in two dimensions $F_{\mathrm{2D}}(\alpha, U, \beta)$ and in three dimensions $F_{\mathrm{3D}}(\alpha, U, \beta)$ [301,302]:

$$F_{\mathrm{2D}}(\alpha, U, \beta) = \frac{2}{3} F_{\mathrm{3D}}\left(\frac{3\pi}{4}\alpha, \frac{3\pi}{4}U, \beta\right). \tag{4.24}$$

According to (4.24), the critical value of the coupling constant for bipolaron formation α_{c} turns out to scale with a factor $3\pi/4 \approx 2.36$ or $\alpha_{\mathrm{c}}^{(\mathrm{2D})} = \alpha_{\mathrm{c}}^{(\mathrm{3D})}/2.36$. From Fig. 4.1b it is seen that bipolarons in 2D can be stable for $\alpha \geq 2.9$, a domain of coupling constants which is definitely realized in several solids. The "characteristic line" $U = 1.526\alpha$ for the material parameters of $\mathrm{YBa_2Cu_3O_7}$ enters the region of bipolaron stability in 2D at a value of α which is appreciably smaller than in the case of $\mathrm{La_2CuO_4}$. This fact suggests $\mathrm{YBa_2Cu_3O_7}$ as a good candidate for the occurrence of stable Fröhlich bipolarons.

An analytical strong-coupling asymptotic expansion in inverse powers of the electron–phonon coupling constant for the large bipolaron energy at $T=0$ was derived in [309]:

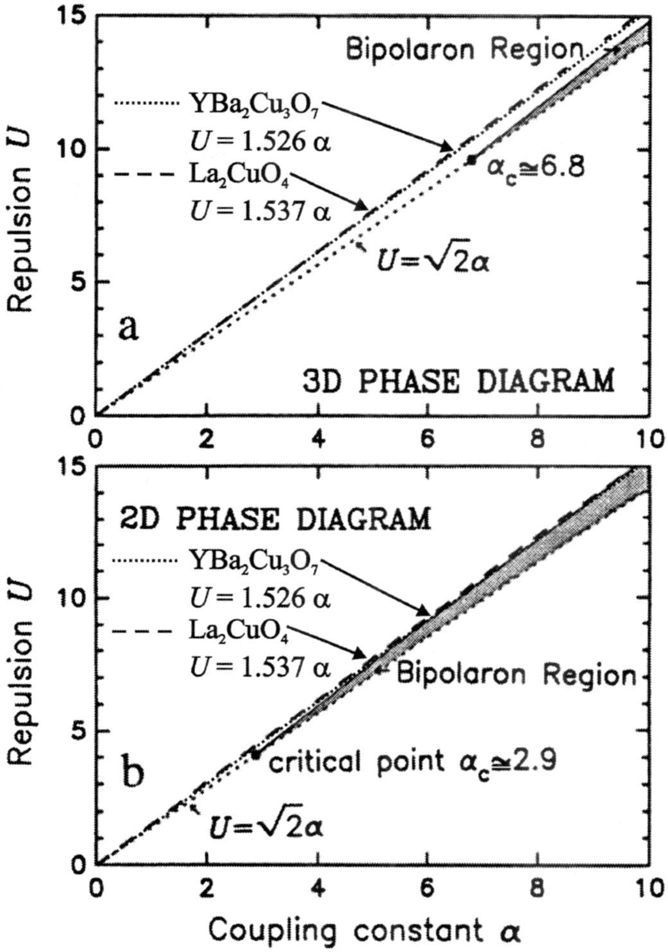

Fig. 4.1. The stability region for bipolaron formation in 3D (**a**) and in 2D (**b**). The *dotted line* $U = \sqrt{2}\alpha$ separates the physical region ($U \geq \sqrt{2}\alpha$) from the non-physical ($U \leq \sqrt{2}\alpha$). The *shaded area* is the stability region in physical space. The *dashed (dotted) characteristic line* $U = 1.537\alpha$ ($U = 1.526\alpha$) is determined by $U = \sqrt{2}\alpha/(1 - \varepsilon/\varepsilon_0)$ where we took the experimental values $\varepsilon = 4$ and $\varepsilon_0 = 50$ for La_2CuO_4 ($\varepsilon = 4.7$ [307] and $\varepsilon_0 = 64.7$ calculated using the experimental data of [307, 308] for $YBa_2Cu_3O_7$). The critical points $\alpha_c = 6.8$ for 3D and $\alpha_c = 2.9$ for 2D are represented as *full dots* (reprinted with permission after [301]. © 1990 by Elsevier)

$$E_{3D}(\alpha, u) = -\frac{2\alpha^2}{3\pi}A(u) - B(u) + O(\alpha^{-2}), \tag{4.25}$$

where the coefficients are closed analytical functions of the ratio $u = U/\alpha$:

$$A(u) = 4 - 2\sqrt{2}u\left(1 + \frac{u^2}{128}\right)^{3/2} + \frac{5}{8}u^2 - \frac{u^4}{512} \tag{4.26}$$

and for $B(u)$ see the above-cited paper. The scaling relation (4.24) allows one to find the bipolaron energy in two dimensions as well.

The stability of bipolarons has also been examined with the use of operator techniques with a variational approach [296]. The bipolaron is bound if the electron–phonon coupling constant α is larger than $\sim$6 in three dimensions and larger than $\sim$2 in two dimensions, provided the ratio $\eta = \varepsilon/\varepsilon_0$ is smaller than a critical value η_c which depends on α. The critical value η_c is larger in the two-dimensional case than in the three-dimensional one. The bipolaron radius is shown to be of the order of a few polaron radii. The results of [296] and [301,302] tend to qualitatively confirm each other. Furthermore, bipolaron states obtained in [296] under the assumption that the total linear momentum is conserved, have intrinsically high mobility.

In the framework of the renewed interest in bipolaron theory after the discovery of high-T_c superconductivity, an analysis of the optical absorption by large [247, 310] and small [281] bipolarons was given.

4.4 Discrete Superlight Fröhlich Bipolaron

The Holstein model is an extreme polaron model, and typically yields the highest possible values of the (bi)polaron mass in the strong-coupling regime (except the case when the lattice vibrations are polarized along the hopping direction [166]). Many doped ionic lattices are characterized by poor screening of high-frequency optical phonons and they are more appropriately described by the finite-range Fröhlich EPI [59]. The unscreened Fröhlich EPI provides relatively light lattice polarons (3.3) and also "superlight" small bipolarons, which are several orders of magnitude lighter than bipolarons in HHM [59, 311, 312].

To illustrate the point, let us consider a generic "Fröhlich–Coulomb" model (FCM) on a lattice, which explicitly includes the finite-range Coulomb repulsion and the strong *long-range* EPI [59, 311]. The implicitly present (infinite) Hubbard U prohibits double occupancy and removes the need to distinguish the fermionic spin, if we are interested in the charge rather than spin excitations. Introducing spinless fermion annihilation operators $c_{\mathbf{n}}$ and phonon annihilation operators $d_{\mathbf{m}}$, the Hamiltonian of FCM is written in the real-space representation as [311]

$$\begin{aligned} H = & \sum_{\mathbf{n}\neq\mathbf{n}'} T(\mathbf{n}-\mathbf{n}')c_{\mathbf{n}}^{\dagger}c_{\mathbf{n}'} + \frac{1}{2}\sum_{\mathbf{n}\neq\mathbf{n}'} V_c(\mathbf{n}-\mathbf{n}')c_{\mathbf{n}}^{\dagger}c_{\mathbf{n}}c_{\mathbf{n}'}^{\dagger}c_{\mathbf{n}'} \\ & + \omega_0 \sum_{\mathbf{n}\neq\mathbf{m}} g(\mathbf{m}-\mathbf{n})(\mathbf{e}\cdot\mathbf{e}_{\mathbf{m}-\mathbf{n}})c_{\mathbf{n}}^{\dagger}c_{\mathbf{n}}(d_{\mathbf{m}}^{\dagger}+d_{\mathbf{m}}) + \omega_0\sum_{\mathbf{m}}\left(d_{\mathbf{m}}^{\dagger}d_{\mathbf{m}}+\frac{1}{2}\right), \end{aligned} \tag{4.27}$$

where $T(\mathbf{n})$ is the bare hopping integral in a rigid lattice.

If we are interested in the non- or near-adiabatic limit and the strong EPI, the kinetic energy is a perturbation. Then, the model can be grossly simplified using the Lang–Firsov canonical transformation. In particular, lattice structures like a staggered triangular ladder in Fig. 4.3 the intersite lattice bipolarons tunnel already in the first order in $t(\mathbf{n})$. That allows us to average the transformed Hamiltonian over phonons to obtain its polaronic part as $H_{\mathrm{p}} = H_0 + H_{\mathrm{pert}}$, where

$$H_0 = -E_{\mathrm{p}} \sum_{\mathbf{n}} c_{\mathbf{n}}^{\dagger} c_{\mathbf{n}} + \frac{1}{2} \sum_{\mathbf{n} \neq \mathbf{n}'} v(\mathbf{n}-\mathbf{n}') c_{\mathbf{n}}^{\dagger} c_{\mathbf{n}} c_{\mathbf{n}'}^{\dagger} c_{\mathbf{n}'},$$

and

$$H_{\mathrm{pert}} = \sum_{\mathbf{n} \neq \mathbf{n}'} t(\mathbf{n}-\mathbf{n}') c_{\mathbf{n}}^{\dagger} c_{\mathbf{n}'}$$

is a perturbation. E_{p} is the familiar polaron level shift:

$$E_{\mathrm{p}} = \omega \sum_{\mathbf{m}} g^2(\mathbf{m}-\mathbf{n})(\mathbf{e} \cdot \mathbf{e}_{\mathbf{m}-\mathbf{n}})^2, \tag{4.28}$$

which is independent of $\mathbf{n}$. The polaron–polaron interaction is

$$v(\mathbf{n}-\mathbf{n}') = V_{\mathrm{c}}(\mathbf{n}-\mathbf{n}') - V_{\mathrm{ph}}(\mathbf{n}-\mathbf{n}'), \tag{4.29}$$

where

$$V_{\mathrm{ph}}(\mathbf{n}-\mathbf{n}') = 2\omega_0 \sum_{\mathbf{m}} g(\mathbf{m}-\mathbf{n}) g(\mathbf{m}-\mathbf{n}')(\mathbf{e} \cdot \mathbf{e}_{\mathbf{m}-\mathbf{n}})(\mathbf{e} \cdot \mathbf{e}_{\mathbf{m}-\mathbf{n}'}). \tag{4.30}$$

The transformed hopping integral is $t(\mathbf{n}-\mathbf{n}') = T(\mathbf{n}-\mathbf{n}') \exp[-g^2(\mathbf{n}-\mathbf{n}')]$ with

$$g^2(\mathbf{n}-\mathbf{n}') = \sum_{\mathbf{m}} g(\mathbf{m}-\mathbf{n})(\mathbf{e}.\mathbf{e}_{\mathbf{m}-\mathbf{n}}) \times [g(\mathbf{m}-\mathbf{n})(\mathbf{e} \cdot \mathbf{e}_{\mathbf{m}-\mathbf{n}}) - g(\mathbf{m}-\mathbf{n}')(\mathbf{e} \cdot \mathbf{e}_{\mathbf{m}-\mathbf{n}'})] \tag{4.31}$$

at $T \ll \omega_0$. The mass renormalization exponent can be expressed via E_{p} and V_{ph} as

$$g^2(\mathbf{n}-\mathbf{n}') = \frac{1}{\omega_0} \left[E_{\mathrm{p}} - \frac{1}{2} V_{\mathrm{ph}}(\mathbf{n}-\mathbf{n}') \right]. \tag{4.32}$$

When V_{ph} exceeds V_{c} the full interaction becomes negative and polarons form pairs. The real-space representation allows us to elaborate more physics behind the lattice sums in V_{ph} [311]. When a carrier (electron or hole) acts on an ion with a force $\mathbf{f}$, it displaces the ion by some vector $\mathbf{x} = \mathbf{f}/k$. Here k is the ion's force constant. The total energy of the carrier-ion pair is $-\mathbf{f}^2/(2k)$.

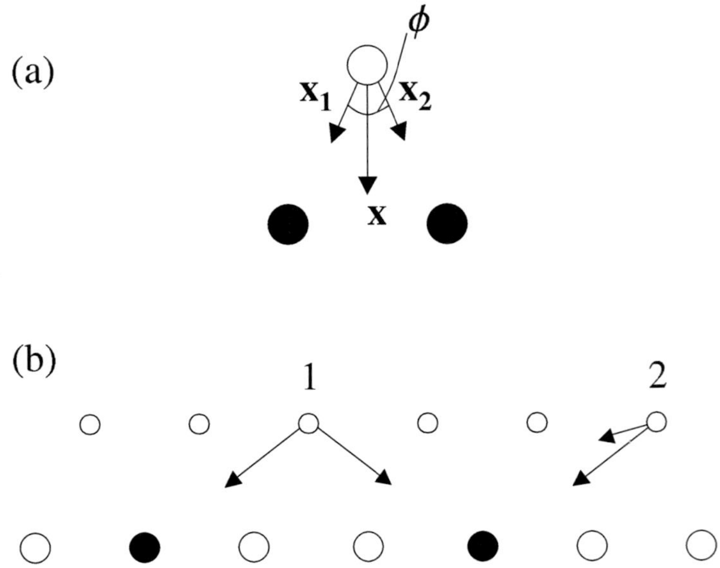

Fig. 4.2. Mechanism of the polaron–polaron interaction. (**a**) Together, two polarons (*solid circles*) deform the lattice more effectively than separately. An effective attraction occurs when the angle ϕ between $\mathbf{x}_1$ and $\mathbf{x}_2$ is less than $\pi/2$. (**b**) A mixed situation: ion 1 results in repulsion between two polarons while ion 2 results in attraction (after [311])

This is precisely the summand in (4.28) expressed via dimensionless coupling constants. Now, consider two carriers interacting with the *same* ion. The ion displacement is $\mathbf{x} = (\mathbf{f}_1+\mathbf{f}_2)/k$ and the energy is $-\mathbf{f}_1^2/(2k)-\mathbf{f}_2^2/(2k)-(\mathbf{f}_1\cdot\mathbf{f}_2)/k$. Here, the last term should be interpreted as an ion-mediated interaction between the two carriers. It depends on the scalar product of $\mathbf{f}_1$ and $\mathbf{f}_2$ and consequently on the relative positions of the carriers with respect to the ion. If the ion is an isotropic harmonic oscillator, as we assume here, then the following simple rule applies. If the angle ϕ between $\mathbf{f}_1$ and $\mathbf{f}_2$ is less than $\pi/2$, the polaron–polaron interaction will be attractive, otherwise it will be repulsive. In general, some ions will generate attraction, and some ions – repulsion between polarons (Fig. 4.2).

The overall sign and magnitude of the interaction is given by the lattice sum in (4.30). One should note that according to (4.32), an attractive EPI reduces the polaron mass (and consequently the bipolaron mass), while repulsive EPI enhances the mass. Thus, the long-range EPI serves a double purpose. Firstly, it generates an additional interpolaron attraction because the distant ions have a small angle ϕ. This additional attraction helps to overcome the direct Coulomb repulsion between polarons. And secondly, the Fröhlich EPI makes lattice bipolarons lighter.

The many-particle ground state of H_0 depends on the sign of the polaron–polaron interaction, the carrier density, and the lattice structure. Following [311], we consider the staggered ladder (Fig. 4.3), assuming that all sites are isotropic two-dimensional harmonic oscillators. For simplicity, we also adopt the nearest-neighbor approximation for both interactions, $g(\mathbf{l}) \equiv g$, $V_\mathrm{c}(\mathbf{n}) \equiv V_\mathrm{c}$, and for the hopping integrals, $T(\mathbf{m}) = T(a)$ for $l = n = m = a$, and zero otherwise. Hereafter, we set the lattice period $a = 1$. There are four nearest neighbors in the ladder, $z = 4$. The *single-particle* polaronic Hamiltonian takes the form

$$H_\mathrm{p} = -E_\mathrm{p} \sum_n (c_n^\dagger c_n + p_n^\dagger p_n) \tag{4.33}$$

$$+ \sum_n [t'(c_{n+1}^\dagger c_n + p_{n+1}^\dagger p_n) + t(p_n^\dagger c_n + p_{n-1}^\dagger c_n) + \mathrm{H.c.}],$$

where c_n and p_n are polaron annihilation operators on the lower and upper legs of the ladder, respectively (Fig. 4.3). Using (4.28), (4.30), and (4.32), one obtains $E_\mathrm{p} = 4g^2\omega_0$, $t' = T(a)\exp(-7E_\mathrm{p}/8\omega_0)$, and $t = T(a)\exp(-3E_\mathrm{p}/4\omega_0)$.

The Fourier transform of H_p yields two overlapping polaron bands:

$$E_\mathrm{p}(k) = -E_\mathrm{p} + 2t'\cos(k) \pm 2t\cos(k/2) \tag{4.34}$$

with the effective mass $m^* = 2/|4t' \pm t|$ near their edges.

Let us now place two polarons on the ladder. The nearest-neighbor interaction is $v = V_\mathrm{c} - E_\mathrm{p}/2$, if two polarons are on different legs of the ladder, and $v = V_\mathrm{c} - E_\mathrm{p}/4$, if both polarons are on the same leg. The attractive interaction is provided via the displacement of the lattice sites, which are the common nearest neighbors to both polarons. There are two such nearest neighbors for the intersite bipolaron of type A or B (Fig. 4.3c), but there is only one common nearest neighbor for bipolaron C (Fig. 4.3d). When $V_\mathrm{c} > E_\mathrm{p}/2$, there are no bound states and the multipolaron system is a one-dimensional Luttinger liquid. However, when $V_\mathrm{c} < E_\mathrm{p}/2$ and consequently $v < 0$, the two polarons are bound into an intersite bipolaron of types A or B.

Remarkably, bipolarons tunnel in the ladder already in the first order with respect to the single-polaron tunneling amplitude. This case is different both from on-site bipolarons discussed above and from intersite chain bipolarons of [313], where the intersite bipolaron tunneling appeared in the second order in t as for the on-site bipolarons. Indeed, the lowest-energy configurations A and B are degenerate. They are coupled by H_pert. Neglecting all higher-energy configurations, we can project the Hamiltonian onto the subspace containing A, B, and empty sites. The result of such a projection is the bipolaronic Hamiltonian

$$H_\mathrm{b} = \left(V_\mathrm{c} - \frac{5}{2}E_\mathrm{p}\right) \sum_n [A_n^\dagger A_n + B_n^\dagger B_n] - t' \sum_n [B_n^\dagger A_n + B_{n-1}^\dagger A_n + \mathrm{H.c.}], \tag{4.35}$$

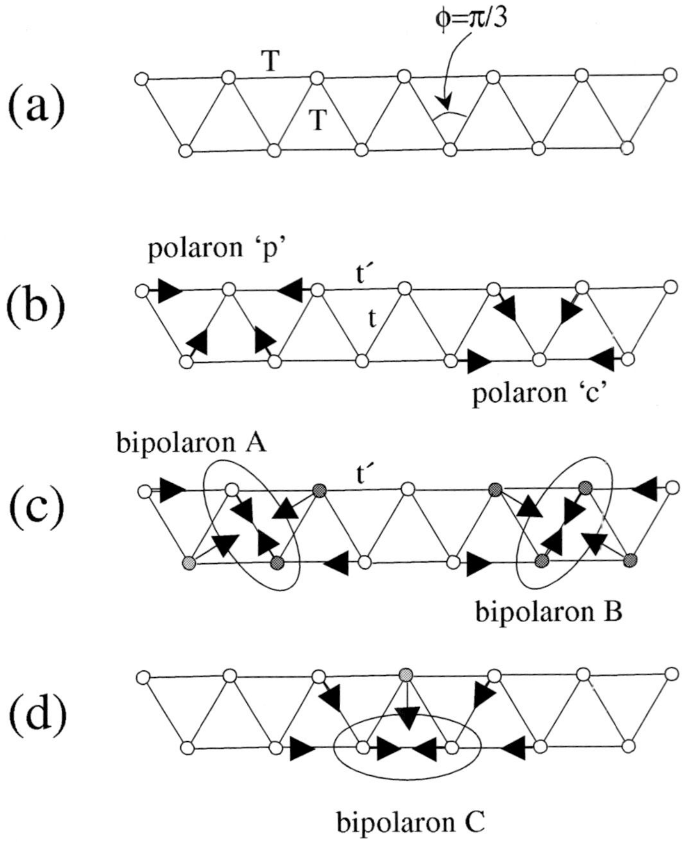

Fig. 4.3. One-dimensional zigzag ladder. (**a**) Initial ladder with the bare hopping amplitude $T(a)$. (**b**) Two types of polarons with their respective deformations. (**c**) Two degenerate bipolaron configurations A and B (here t, t' are renormalized (i.e., polaronic) hopping integrals). (**d**) A different bipolaron configuration, C, whose energy is higher than that of A and B

where $A_n = c_n p_n$ and $B_n = p_n c_{n+1}$ are intersite bipolaron annihilation operators, and the bipolaron–bipolaron interaction is omitted. The Fourier transform of (4.35) yields two *bipolaron* bands:

$$E_2(k) = V_{\rm c} - \frac{5}{2}E_{\rm p} \pm 2t' \cos(k/2) \tag{4.36}$$

with a combined width $4|t'|$. The bipolaron binding energy in zero order with respect to t, t' is $\Delta \equiv 2E_1(0) - E_2(0) = E_{\rm p}/2 - V_{\rm c}$.

The bipolaron mass near the bottom of the lowest band, $m^{**} = 2/t'$, is

$$m^{**} = 4m^* \left[1 + 0.25 \exp\left(\frac{E_{\rm p}}{8\omega_0}\right)\right]. \tag{4.37}$$

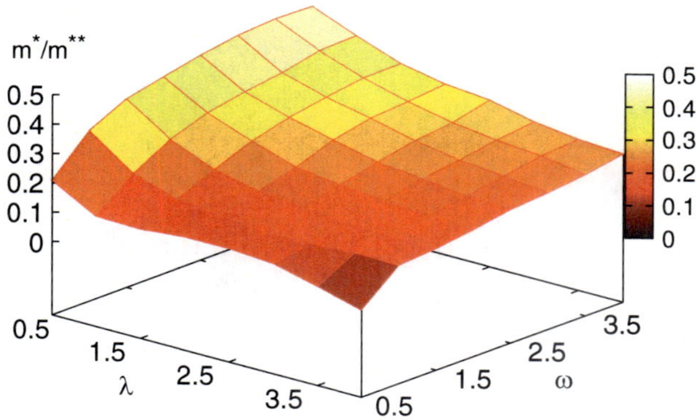

Fig. 4.4. Polaron to bipolaron mass ratio for a range of $\bar{\omega} = \omega_0/T(a)$ and λ on the staggered ladder. Mobile small bipolarons are seen even in the adiabatic regime $\bar{\omega} = 0.5$ for couplings λ up to 2.5 (reproduced from [312], © American Physical Society, 2007)

The numerical coefficient 1/8 in the exponent ensures that m^{**} remains of the order of m^* even at sufficiently large $E_{\rm p}$ up to $E_{\rm p} \approx 10\omega_0$. This fact combined with a weaker renormalization of m^* provides a *superlight* small bipolaron [59, 311, 312].

4.5 Discrete All-Coupling Fröhlich Bipolaron

The FCM model discussed above is analytically solvable in the strong-coupling nonadiabatic ($\omega_0 \gtrsim T(a)$) limit using the Lang–Firsov transformation of the Hamiltonian, and projecting it on the intersite pair Hilbert space [59, 311]. The theory has been extended to the whole parameter space using CTQMC technique for bipolarons [312]. Hague *et al.* [312] simulated the FCM Hamiltonian on a staggered triangular ladder (1D), triangular (2D), and strongly anisotropic hexagonal (3D) lattices including triplet pairing. On such lattices, bipolarons are found to move with a crab-like motion (Fig. 4.3), which is distinct from the crawler motion found on cubic lattices [271]. Such bipolarons are small but very light for a wide range of electron–phonon couplings and phonon frequencies. EPI has been modeled using the force function in the site representation as in (3.63). Coulomb repulsion has been screened up to the first nearest neighbors, with on-site repulsion U and nearest-neighbor repulsion $V_{\rm c}$. The dimensionless electron–phonon coupling constant λ is defined as $\lambda = \sum_{\mathbf{m}} f^2_{\mathbf{m}}(0)/2M\omega^2 zT(a)$ which is the ratio of the polaron binding energy to the kinetic energy of the free electron $zT(a)$, and the lattice constant is taken as $a = 1$.

Extending the CTQMC algorithm to systems of two particles with strong EPI and Coulomb repulsion solved the bipolaron problem on different

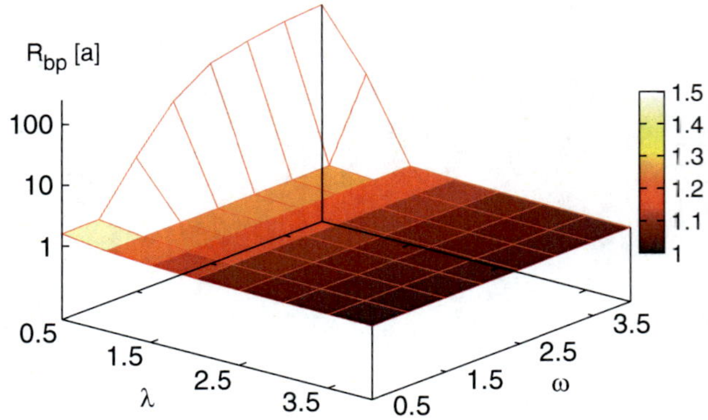

Fig. 4.5. Bipolaron radius (in units of a) for a range of $\bar{\omega}$ and λ on the staggered ladder (reproduced from [312], © American Physical Society, 2007)

lattices from weak to strong coupling in a realistic parameter range where usual strong- and weak-coupling limiting approximations fail.

Figure 4.4 shows the ratio of the polaron to bipolaron masses on the staggered ladder as a function of effective coupling and phonon frequency for $V_c = 0$. The bipolaron to polaron mass ratio is about 2 in the weak-coupling regime ($\lambda \ll 1$) as it should be for a large bipolaron (Sect. 4.3). In the strong-coupling, large phonon frequency limit the mass ratio approaches 4, in agreement with strong-coupling results (Sect. 4.4). In a wide region of parameter space, the bipolaron/polaron mass ratio has been found between 2 and 4 and a bipolaron radius similar to the lattice spacing (see Figs. 4.4 and 4.5). Thus, the bipolaron is small and light at the same time. Taking into account additional intersite Coulomb repulsion V_c does not change this conclusion. The bipolaron is stable for $V_c < 4T(a)$. As V_c increases the bipolaron mass decreases but the radius remains small, at about two lattice spacings. Importantly, the absolute value of the small bipolaron mass is only about four times of the bare-electron mass m_0, for $\lambda = \omega/T(a) = 1$ (see Fig. 4.4).

Simulations of the bipolaron on an infinite triangular lattice including exchanges and large on-site Hubbard repulsion $U = 20T(a)$ also lead to the bipolaron mass of about $6m_{0xy}$ and the bipolaron radius $R_{\mathrm{bp}} \approx 2a$ for a moderate coupling $\lambda = 0.5$ and a large phonon frequency $\omega = T(a)$ (for the triangular lattice, $m_{0xy} = 1/3a^2T(a)$). Finally, the bipolaron in a hexagonal lattice with out-of-plane hopping $T' = T(a)/3$ has also a light in-plane inverse mass, $m_{xy}^{**} \approx 4.5m_{0xy}$ but a small size, $R_{\mathrm{bp}} \approx 2.6a$ for experimentally achievable values of the phonon frequency $\omega = T(a) = 200\,\mathrm{meV}$ and EPI, $\lambda = 0.36$. Out-of-plane $m_z^{**} \approx 70m_{0z}$ is Holstein like, where $m_{0z} = 1/2d^2T'$ (d is the interplane spacing). When bipolarons are small and pairs do not overlap, the pairs can form a Bose-Einstein condensate (BEC) at $T_{\mathrm{BEC}} = 3.31$ $(2n_{\mathrm{B}}/a^2\sqrt{3}d)^{2/3}/(m_{xy}^{2/3}m_z^{1/3})$. If we choose realistic values for the lattice constants of 0.4 nm in the plane and 0.8 nm out of the plane, and allow the density of bosons to be $n_{\mathrm{B}} = 0.12$ per lattice site, which easily avoids overlap of pairs, then $T_{\mathrm{BEC}} \approx 300\,\mathrm{K}$.

4.6 Polaronic Exciton

We also briefly mention works on electron–hole bound states coupled with phonons. Such excitonic polaron states have been analyzed by Iadonisi and Bassani [314–316]. The binding energies of excitonic states in interaction with LO phonons were computed using a phonon coherent state and applying a variational method [314]. Incomplete relaxation of the lattice is found. The binding energies are larger than those obtained with static dielectric screening when the polaron radius is comparable to the exciton radius. Analyzing the relative intensities of the one-phonon and zero-phonon lines for a number of semiconductors, Iadonisi and Bassani [314] observed that the zero-phonon exciton states are generally much more probable than the phonon replicas.

The binding energy of a core exciton, which depends on the interaction of the conduction electron and of the core hole with the valence electrons via a Fröhlich-type coupling with the electron–hole pairs, in addition to the Coulomb attraction, was calculated within a functional variational method [315]. When the exciton radius is comparable to the polaron radius, the static dielectric screening reduces and the core exciton binding energy increases. The excitonic polaron effective mass renormalization was analyzed by Iadonisi and Bassani [315] using a variational numerical approach. Evidence for this renormalization is found from the experimental data on polariton dispersion in hexagonal CdS and in CuCl.

The angular momentum as a constant of the motion of a Fröhlich polaron was introduced by Evrard *et al.* [317]. Representing the excitonic polaron problem in angular coordinates and solving it numerically, Iadonisi *et al.* [318] showed that exciton states with various total angular momenta L are differently affected by the interaction with LO phonons. However, when extending this approach to the bipolaron problem, the same authors found that the Fröhlich interaction is not strong enough to guarantee a stable bipolaron state at least in the case of cubic materials.

The approximation-free diagrammatic Monte Carlo technique has also been applied to the exciton–polaron problem, providing numerically exact results for the wave function, ground-state energy, binding energy, and effective mass of polaronic excitons [319].

5

Multipolaron Problem

5.1 Ground State of a Large-Polaron Gas, Polaron Wigner Crystals, and Ripplopolarons

For the weak-coupling regime, which is realized in most polar semiconductors, the ground-state energy of a gas of interacting continuous polarons has been derived in [320] by introducing a variational wave function:

$$|\psi_{\mathrm{LDB}}\rangle = U\,|\phi\rangle\,|\varphi_{\mathrm{el}}\rangle\,, \tag{5.1}$$

where $|\varphi_{\mathrm{el}}\rangle$ represents the ground-state many-body wave function for the electron (or hole) system, $|\phi\rangle$ is the phonon vacuum, and U is a many-body unitary operator. U defines the LDB-canonical transformation for a fermion gas interacting with a boson field:

$$U = \exp\left\{\sum_{j=1}^{N}\sum_{\mathbf{q}}\left(f_{\mathbf{q}}d_{\mathbf{q}}\mathrm{e}^{\mathrm{i}\mathbf{q}\cdot\mathbf{r}_j} - f_{\mathbf{q}}^{*}d_{\mathbf{q}}^{+}\mathrm{e}^{-\mathrm{i}\mathbf{q}\cdot\mathbf{r}_j}\right)\right\}, \tag{5.2}$$

where $\mathbf{r}_j$ represent the position of the N constituent electrons (or holes). The $f_{\mathbf{q}}$ were determined variationally [320]. It may be emphasized that (5.2), although it appears like a straightforward generalization of the one-particle transformation in [321], constitutes – especially in its implementation – a nontrivial extension of a one-particle approximation to a many-body system. An advantage of the LDB-many-polaron canonical transformations introduced in [320] for the calculation of the ground-state energy of a polaron gas is that the many-body effects are contained in the static structure factor of the electron (or hole) system, which appears in the analytical expression for the energy. Within the approach, the minimum of the total ground-state energy per particle for a polaron gas lies at lower density than that for the electron gas.

The nondegenerate system of interacting polarons in polar doped insulators was analyzed by Fratini and Quémerais [322] using a simplified Feynman-type polaron model. The phase diagram in Fig. 5.1 presents the mean distances

A.S. Alexandrov and J.T. Devreese, *Advances in Polaron Physics*, Springer Series in Solid-State Sciences 159,
DOI: 10.1007/978-3-642-01896-1_5, © Springer-Verlag Berlin Heidelberg 2010

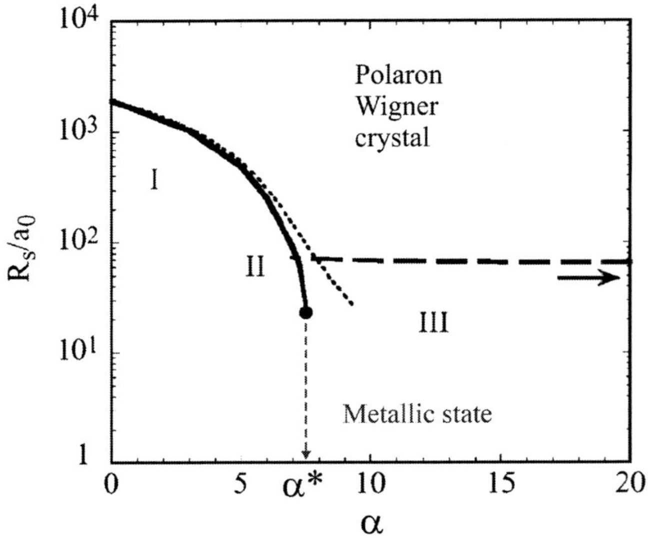

Fig. 5.1. Model for the phase diagram of the polaron Wigner lattice. R_S is the mean distance between two electrons at a given density, $a_0 = 1/m_e e^2$ is the Bohr radius. Crystal melting and polaron dissociation are represented with *bold and broken lines*, respectively. The *dotted line* indicates the mean distance between two electrons corresponding to the low-density approximation. The *arrow* corresponds to the critical density of the polaron dissociation. The melting driven by the fluctuations of the localized particles is impossible beyond α^* (the *filled dot*). For $\alpha < \alpha^*$, the metallic state beyond the critical density changes with increasing α from a weak-coupling polaron liquid (region I) to a strong-coupling polaron liquid (region II). For $\alpha > \alpha^*$, the insulator-to-metal transition is described by the dissociation of the polarons(region III) (reprinted with permission after [322]. © 2000, EDP Sciences, Società Italiana di Fisica, Springer)

between two electrons for Wigner crystal melting and polaron dissociation as a function of the electron–phonon coupling constant α. In the low-density limit, the ground state of the many-polaron system is the Wigner lattice of polarons. With increasing density depending on the value of α, one of the following two scenarios is possible (1) the melting of the polaron Wigner lattice for $\alpha < \alpha^*$ and (2) the dissociation of the polarons for $\alpha > \alpha^*$ [322].

An interesting 2D system consists of electrons on films of liquid He [323, 324]. In this system, the electrons couple to the ripplons of the liquid He, forming "ripplopolarons." The effective coupling can be relatively large and self-trapping can result. The acoustic nature of the ripplon dispersion at long wavelengths induces the self-trapping. Spherical shells of charged particles appear in a variety of physical systems, such as fullerenes, metallic nanoshells, charged droplets, and neutron stars. A particularly interesting physical realization of the spherical electron gas is found in multielectron bubbles (MEBs) in liquid helium-4. These MEBs are 0.1–100 μm sized cavities inside liquid helium that contain helium vapor at vapor pressure and a nanometer-thick electron layer, anchored to the surface of the bubble [325, 326]. They exist as a result of equilibrium between the surface tension of liquid helium and the Coulomb repulsion of the electrons [327, 328].

Recently proposed experimental schemes to stabilize MEBs [329] have stimulated theoretical investigation of their properties (see, e.g., [330]). The dynamical modes of MEB were described by considering the motion of the helium surface ("ripplons") and the vibrational modes of the electrons together. In particular, the case when the ripplopolarons form a Wigner lattice was analyzed in [331]. The interaction energy between the ripplons and the electrons in the multielectron bubble is derived from the following considerations (1) the distance between the layer electrons and the helium surface is fixed (the electrons find themselves confined to an effectively 2D surface anchored to the helium surface) and (2) the electrons are subjected to a force field, arising from the electric field of the other electrons. To study the ripplopolaron Wigner lattice at nonzero temperature and for any value of the electron–ripplon coupling, the variational path-integral approach [12] is used.

In their treatment of the electron Wigner lattice embedded in a polarizable medium, such as a semiconductor or an ionic solid, Fratini and Quémerais [322] described the effect of the electrons on a particular electron through a mean-field lattice potential. The (classical) lattice potential V_{lat} is obtained by approximating all the electrons acting on one particular electron by a homogeneous charge density, in which a hole is punched out; this hole is centered in the lattice point of the particular electron under investigation and has a radius given by the lattice distance d.

The Lindemann melting criterion states in general that a crystal lattice of objects (be it atoms, molecules, electrons, or ripplopolarons) will melt when the average displacement of the objects from their lattice site is larger than a critical fraction δ_0 of the lattice parameter d. It would be a strenuous task to calculate, from first principles, the exact value of the critical fraction δ_0, but for the particular case of electrons on a helium surface, we can make use of an experimental determination. Grimes and Adams [332] found that the Wigner lattice melts when $\Gamma = 137 \pm 15$, where Γ is the ratio of potential energy to the kinetic energy per electron. At temperature T, the average kinetic energy of an electron in a lattice potential V_{lat}, characterized by the frequency parameter ω_{lat}, is

$$E_{\mathrm{kin}} = \frac{\omega_{\mathrm{lat}}}{2} \coth\left(\frac{\omega_{\mathrm{lat}}}{2T}\right), \tag{5.3}$$

and the average distance that an electron moves out of the lattice site is determined by

$$\left\langle \mathbf{r}^2 \right\rangle = \frac{1}{m_{\mathrm{e}}\omega_{\mathrm{lat}}} \coth\left(\frac{\omega_{\mathrm{lat}}}{2T}\right) = \frac{2E_{\mathrm{kin}}}{m_{\mathrm{e}}\omega_{\mathrm{lat}}^2}. \tag{5.4}$$

From this, one finds that for the melting transition in the experiment in [332], the critical fraction equals $\delta_0 \approx 0.13$. This estimate is in agreement with previous (empirical) estimates yielding $\delta_0 \approx 0.1$ [333].

Within the approach of [322], the Wigner lattice of (ripplo)polarons melts when at least one of the two following Lindemann criteria is met:

$$\delta_{\mathrm{r}} = \frac{\sqrt{\langle \mathbf{R}_{\mathrm{cms}}^2 \rangle}}{d} > \delta_0, \tag{5.5}$$

$$\delta_{\rho} = \frac{\sqrt{\langle \rho^2 \rangle}}{d} > \delta_0, \tag{5.6}$$

where ρ and $\mathbf{R}_{\mathrm{cms}}$ are, respectively, the relative coordinate and the center of mass coordinate of the model system: if $\mathbf{r}$ is the electron coordinate and $\mathbf{R}$ is the position coordinate of the fictitious ripplon mass M, this is

$$\mathbf{R}_{\mathrm{cms}} = \frac{m_{\mathrm{e}}\mathbf{r} + M\mathbf{R}}{m_{\mathrm{e}} + M}, \quad \rho = \mathbf{r} - \mathbf{R}. \tag{5.7}$$

The appearance of two Lindemann criteria takes into account the composite nature of (ripplo)polarons. As follows from the physical meaning of the coordinates ρ and $\mathbf{R}_{\mathrm{cms}}$, the first criterion (5.5) is related to the melting of the ripplopolaron Wigner lattice toward a ripplopolaron liquid, where the ripplopolarons move as a whole, the electron together with its dimple. The second criterion (5.6) is related to the dissociation of ripplopolarons: the electrons shed their dimple.

The Feynman path-integral variational formalism allows us to calculate the expectation values $\langle \mathbf{R}_{\mathrm{cms}}^2 \rangle$ and $\langle \rho^2 \rangle$ with respect to the ground state of the variationally optimal model system.

Numerical calculation shows that for ripplopolarons in an MEB, the inequality

$$\langle \mathbf{R}_{\mathrm{cms}}^2 \rangle \ll \langle \rho^2 \rangle \tag{5.8}$$

is realized. As a consequence, the destruction of the ripplopolaron Wigner lattice in an MEB occurs through the dissociation of ripplopolarons, since the second criterion (5.6) will be fulfilled before the first (5.5). The results for the melting of the ripplopolaron Wigner lattice are summarized in the phase diagram shown in Fig. 5.2.

For any value of N, pressure p, and temperature T in an experimentally accessible range, Fig. 5.2 shows whether the ripplopolaron Wigner lattice is present (points above the surface) or the electron liquid (points below the surface). Below a critical pressure (on the order of 10^4 Pa), the ripplopolaron solid will melt into an electron liquid. This critical pressure is nearly independent of the number of electrons (except for the smallest bubbles) and is weakly temperature dependent, up to the helium critical temperature 5.2 K. This can be understood since the typical lattice potential well in which the ripplopolaron resides has frequencies of the order of THz or larger, which correspond to ~10 K.

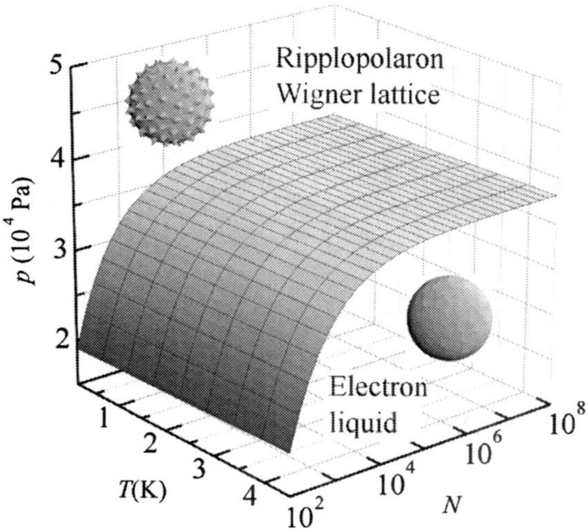

Fig. 5.2. The phase diagram for the spherical 2D layer of electrons in an MEB. Above a critical pressure, a ripplopolaron solid (a Wigner lattice of electrons with dimples in the helium surface underneath them) is formed. Below the critical pressure, the ripplopolaron solid melts into an electron liquid through dissociation of ripplopolarons (reprinted with permission from [331]. © 2003, EDP Sciences, Società Italiana di Fisica, Springer)

The new phase that was predicted in [331], the ripplopolaron Wigner lattice, will not be present for electrons on a flat helium surface. At the values of the pressing field necessary to obtain a strong enough electron–ripplon coupling, the flat helium surface is no longer stable against long-wavelength deformations [334]. Multielectron bubbles, with their different ripplon dispersion and the presence of stabilizing factors such as the energy barrier against fissioning [335], allow for much larger electric fields pressing the electrons against the helium surface. The regime of N, p, and T parameters suitable for the creation of a ripplopolaron Wigner lattice lies within the regime that would be achievable in recently proposed experiments, aimed at stabilizing MEBs [329]. The ripplopolaron Wigner lattice and its melting transition might be detected by spectroscopic techniques [332,336], probing, e.g., the transverse phonon modes of the lattice [337].

Another field, where the LDB-canonical transformation has been fruitfully applied, is the theory of optical absorption spectra of many-polaron systems. In [92], starting from the LDB-many-polaron canonical transformation and the variational many-polaron wave function introduced in [320], the optical absorption coefficient of a many-polaron gas has been derived. The real part of the optical conductivity of the many-polaron system is obtained in an intuitively appealing form, given by (2.95).

This approach to the many-polaron optical absorption allows one to include the many-body effects to order α in terms of the dynamical structure

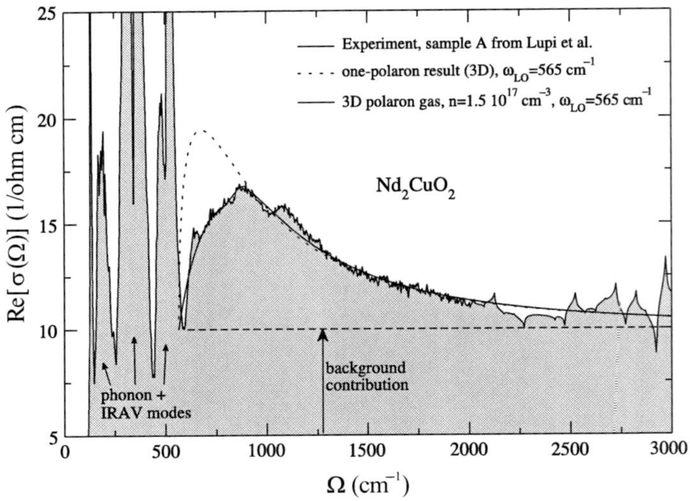

Fig. 5.3. The infrared optical absorption of $Nd_2CuO_{2-\delta}$ ($\delta < 0.004$) as a function of frequency. The experimental results of [338] are presented by the *thin full curve*. The experimental "d-band" is clearly identified, rising in intensity at about 600 cm^{-1}, peaking around 1000 cm^{-1}, and then decreasing in intensity above that frequency. The *dotted curve* shows the single polaron result calculated according to [73]. The *bold full curve* presents the theoretical results of [92] for the interacting many-polaron gas with the following choice of parameters: $n = 1.5 \times 10^{17}$ cm^{-3} (polaron density), $\alpha = 2.1$, and $m = 0.5m_e$ (reprinted with permission from [92]. © 2001 by the American Physical Society)

factor $S(\mathbf{k}, \Omega - \omega_{LO})$ of the electron (or hole) system. The experimental peaks in the mid-infrared optical absorption spectra of cuprates [338] (Fig. 5.3) and manganites [339] have been adequately interpreted within this theory. The many-polaron approach describes the experimental optical conductivity better than the single-polaron approximations [91,247]. Note that in [92], like in [104], coexistence of small and Fröhlich polarons in the same solid is involved.

Recent experimental observations of the optical conductivity in the Nb-doped $SrTiO_3$ [340] reveal the evidence of the mid-infrared optical conductivity band provided by the polaron mechanism. The effective mass of the charge carriers is obtained by analyzing the Drude spectral weight. Defining the mass renormalization of the charge carriers as the ratio of the total electronic spectral weight and the Drude spectral weight, a twofold mass enhancement is obtained, which is caused by the electron–phonon coupling. The missing spectral weight is recovered according to the sum rule [118] in a mid-infrared optical conductivity band. This band results from the electron–phonon coupling interaction, traditionally associated with the polaronic nature of the charge carriers. The effective mass obtained from the optical spectral weights yields an intermediate electron–phonon coupling strength, $3 < \alpha < 4$.

Therefore, it has been suggested in [340] that the charge transport in the Nb-doped $SrTiO_3$ is carried by large polarons.

The optical conductivity of a many-polaron gas was investigated in [341] in a different way by calculating the correction to the dielectric function of the electron gas, due to the electron–phonon interaction with variational parameters of a single-polaron Feynman model. A suppression of the optical absorption of a many-polaron gas as compared to the one-polaron optical absorption of [73, 93] with increasing density has been found. Such a suppression is expected because of the screening of the Fröhlich interaction with increasing polaron density.

5.2 Breakdown of the Migdal–Eliashberg Theory in the Strong-Coupling Regime

While the single- and two-polaron problems have been actively researched for a long time, the multipolaron physics has gained particular attention in the last two decades. For weak electron–phonon coupling, $\lambda < 1$, and the adiabatic limit, $\omega_0/E_F \ll 1$, Migdal theory describes electron dynamics in the normal Fermi-liquid state [264], and BCS–Eliashberg theory in the superconducting state [263, 342]. If EPI is weak Migdal's theorem is perfectly applicable. The theorem proves that the contribution of diagrams with "crossing" phonon lines (so-called "vertex" corrections) is small if the parameter $\lambda\omega/E_F$ is small, where E_F is the Fermi energy. Neglecting the vertex corrections, Migdal [264] calculated the renormalized electron mass as $m^* = m(1+\lambda)$ (near the Fermi level) and Eliashberg [263] extended Migdal's theory to describe the BCS superconducting state at intermediate values of $\lambda < 1$. Later on many researchers applied Migdal–Eliashberg theory (MET) with λ even larger than 1 (e.g., in [343], and references therein).

Naturally, with increasing strength of EPI and increasing phonon frequency, ω_0, finite bandwidth [344, 345], and vertex corrections [346, 347] become more and more important. But unexpectedly for many researchers, who applied MET even at $\lambda > 1$ with or without vertex corrections, the theory breaks down entirely at $\lambda \sim 1$ for *any* value of the adiabatic ratio ω_0/E_F [130]. It happens since the bandwidth is narrowed (Sect. 3.2.1) and the Fermi energy, E_F, is renormalized down exponentially, so that the effective parameter $\lambda\omega_0/E_F$ becomes large already at $\lambda \simeq 1$. Under certain conditions [125, 130], the multipolaron system is metallic but with polaronic carriers rather than bare electrons. This regime is beyond MET, where the effective mass approximation is used and the electron bandwidth is infinite. In recent years, a number of numerical and analytical studies have confirmed the conclusion [130] that MET breaks down at $\lambda \geq 1$ (see, e.g., [165, 189, 190, 224, 251, 274, 277, 288, 348, 349], and references therein). Strong correlation between polarons shrinks the region of applicability of MET even further [35, 58].

It is easy to illustrate the breakdown using the simple molecular Holstein model in the extreme adiabatic limit, $\omega_0/t \to 0$ [125], where MET is supposed to be exact. The Hamiltonian of the model is

$$H = -t\sum_{\langle ij\rangle} c_i^\dagger c_j + \text{H.c.} + 2(\lambda kt)^{1/2}\sum_i x_i c_i^\dagger c_i + \sum_i\left(-\frac{1}{2M}\frac{\partial^2}{\partial x_i^2} + \frac{kx_i^2}{2}\right), \tag{5.9}$$

where x_i is the normal coordinate of the molecule (site) i, $k = M\omega_0^2$, and the kinetic energy of ions is neglected. In the framework of MET, one would expect the Fermi-liquid behavior above T_c and the BCS ground state below T_c at any value of λ. In fact, the exact ground state is an insulator due to self-trapping at any filling of the band, if $\lambda \geq 1$.

Let us first consider a two-site case (zero-dimensional limit), $i, j = 1, 2$ with one electron. The transformation $X = (x_1 + x_2)$, $\xi = x_1 - x_2$ allows us to eliminate the coordinate X, which is coupled only with the total density ($n_1 + n_2 = 1$). That leaves the following Hamiltonian to be solved in the extreme adiabatic limit $M \to \infty$:

$$H = -t(c_1^\dagger c_2 + c_2^\dagger c_1) + (\lambda kt)^{1/2}\xi(c_1^\dagger c_1 - c_2^\dagger c_2) + \frac{k\xi^2}{4}. \tag{5.10}$$

The solution is

$$\psi = (\alpha c_1^\dagger + \beta c_2^\dagger)\,|0\rangle\,, \tag{5.11}$$

where

$$\alpha = \frac{t}{\left[t^2 + \left((\lambda kt)^{1/2}\xi + (t^2 + \lambda kt\xi^2)^{1/2}\right)^2\right]^{1/2}}, \tag{5.12}$$

$$\beta = -\frac{(\lambda kt)^{1/2}\xi + (t^2 + \lambda kt\xi^2)^{1/2}}{\left[t^2 + \left((\lambda kt)^{1/2}\xi + (t^2 + \lambda kt\xi^2)^{1/2}\right)^2\right]^{1/2}}, \tag{5.13}$$

and the energy is

$$E = \frac{k\xi^2}{4} - (t^2 + \lambda kt\xi^2)^{1/2}. \tag{5.14}$$

In the extreme adiabatic limit, the displacement ξ is classical, so the ground-state energy E_0 and the ground-state displacement ξ_0 are obtained by minimizing (5.14) with respect to ξ. If $\lambda \geq 0.5$, one obtains

$$E_0 = -t\left(\lambda + \frac{1}{4\lambda}\right) \tag{5.15}$$

and

$$\xi_0 = \left[\frac{t(4\lambda^2-1)}{\lambda k}\right]^{1/2}. \tag{5.16}$$

The symmetry-breaking "order" parameter is

$$\Delta \equiv \beta^2 - \alpha^2 = \frac{[2\lambda + (4\lambda^2-1)^{1/2}]^2 - 1}{[2\lambda + (4\lambda^2-1)^{1/2}]^2 + 1}. \tag{5.17}$$

If $\lambda < 0.5$, the ground state is "translation" invariant, $\Delta = 0$, and $E_0 = -t$, $\xi = 0$, $\beta = -\alpha$. However, when $\lambda > 0.5$, this solution is *not* the ground state of the system. The system collapses into a localized adiabatic polaron trapped on the "right-hand" or on the "left-hand" site due to the finite-lattice deformation, $\xi_0 \neq 0$.

The generalization for a multipolaron system on the infinite lattice of any dimension is straightforward in the extreme adiabatic regime. According to Kabanov and Mashtakov [350] the self-trapping of a single electron occurs at $\lambda \geq 0.875$ and at $\lambda \geq 0.92$ in 2D and 3D Holstein model. The radius of the self-trapped adiabatic polaron is about the lattice constant. Hence, the multipolaron system remains in the self-trapped insulating state in the strong-coupling adiabatic regime, no matter how many polarons it has. The only instability which might occur in this regime is the formation of self-trapped on-site bipolarons, if the on-site attractive interaction, $2\lambda zt$, is larger than the repulsive Hubbard U [273,276]. Self-trapped on-site bipolarons form a charge-ordered state due to a weak repulsion between them [271,277,283].

In general, the transition into the self-trapped state due to a broken translational symmetry is expected at $0.5 < \lambda < 1.3$ (depending on the lattice dimensionality) for any EPI conserving the on-site electron occupation numbers. For example, Hiramoto and Toyozawa [297] calculated the strength of the deformation potential, which transforms electrons into small polarons and bipolarons. They found that the transition of two electrons into a self-trapped small bipolaron occurs at the electron-acoustic phonon coupling $\lambda \simeq 0.5$, that is, half of the critical value of λ at which the transition of the electron into the small acoustic polaron takes place in the extreme adiabatic limit. The radius of the acoustic polaron and bipolaron is about the lattice constant, so that the critical value of λ does not depend very much on the number of electrons in this case either. Of course, the nonadiabatic corrections (i.e., quantum phonons) allow small bipolarons to propagate as the Bloch states in narrow bands [271].

Carriers in the fascinating advanced materials are strongly coupled with high-frequency optical phonons, making small polarons and nonadiabatic effects relevant for high-temperature superconductivity, colossal magnetoresistance phenomenon, and molecular electronic devices. Indeed, the characteristic phonon energies 0.05–0.2 eV in cuprates, manganites, and in many organic materials are of the same order as generally accepted values of the hopping integrals, $t \approx 0.1$–0.4 eV.

5.3 Polaronic Superconductivity

The polaron–polaron interaction is the sum of two large contributions of the opposite sign (3.40). It is generally larger than the polaron bandwidth and the polaron Fermi energy, $\epsilon_F = Z' E_F$ [130]. This condition is opposite to the weak-coupling BCS regime, where the Fermi energy is the largest. However, there is still a narrow window of parameters, where pairs of two lattice polarons are overlapped similarly to the Cooper pairs. Here, the BCS approach is applied to nonadiabatic carriers with a *nonretarded* attraction, so that *small polarons* form the *polaronic* Cooper pairs [130], rather than the real-space bipolarons. The size of the pair is estimated as

$$r_{\mathrm{b}} \approx \frac{1}{(m^*\Delta)^{1/2}}, \tag{5.18}$$

where Δ is the binding energy of the order of the attraction potential v. The BCS approach is applied if $r_{\mathrm{b}} \gg a n_{\mathrm{p}}^{-1/3}$, which puts a severe constraint on the value of the attraction:

$$|v| \ll \epsilon_{\mathrm{F}}. \tag{5.19}$$

There is no "Tolmachev–Morel–Anderson" logarithm (see, e.g., in [343]) in the case of nonadiabatic carriers, because the attraction is nonretarded as soon as $\epsilon_{\mathrm{F}} \lesssim \omega_0$. Hence, a superconducting state of lattice polarons is possible only if $\lambda > \mu_{\mathrm{c}}$. This consideration leaves a rather narrow *crossover* region from the normal polaron liquid to a superconductor, where one can still apply the BCS mean-field approach, $0 < \lambda - \mu_{\mathrm{c}} \ll Z' < 1$. In the case of the Fröhlich interaction, Z' is about $0.1 \div 0.3$ for typical values of λ. Hence, this region is on the borderline from the polaronic normal metal to a bipolaronic superconductor (Sect. 5.5).

When EPI is strong, $\lambda > 1$, and the Coulomb repulsion nearly compensates the attraction, small polarons behave like fermions in a narrow band with the weak nonretarded attraction described by the transformed Hamiltonian:

$$\tilde{H} \approx \sum_{i,j} \left[(\langle \hat{\sigma}_{ij} \rangle - \mu \delta_{ij}) c_i^\dagger c_j + \frac{1}{2} v_{ij} c_i^\dagger c_j^\dagger c_j c_i \right] \tag{5.20}$$

in the Wannier representation. If the condition (5.19) is satisfied, we can treat the polaron–polaron interaction approximately using the BCS-like mean-field theory [130]. For simplicity, we can keep only the on-site v_0 and the nearest-neighbor v_1 interactions. At least one of them should be attractive to ensure that the ground state is superconducting. Introducing $\Delta_0 = -v_0 \langle c_{\mathbf{m},\uparrow} c_{\mathbf{m},\downarrow} \rangle$ and $\Delta_1 = -v_1 \langle c_{\mathbf{m},\uparrow} c_{\mathbf{m}+\mathbf{a},\downarrow} \rangle$ and transforming into $\mathbf{k}$-space results in the familiar BCS Hamiltonian:

$$H_{\mathrm{p}} = \sum_{\mathbf{k},s} \xi_{\mathbf{k}} c_{\mathbf{k}s}^\dagger c_{\mathbf{k}s} + \sum_{\mathbf{k}} [\Delta_{\mathbf{k}} c_{\mathbf{k}\uparrow}^\dagger c_{-\mathbf{k}\downarrow}^\dagger + \mathrm{H.c.}], \tag{5.21}$$

where $\xi_{\mathbf{k}} = \epsilon_{\mathbf{k}} - \mu$ is the renormalized kinetic energy and $\Delta_{\mathbf{k}} = \Delta_0 - \Delta_1((\xi_{\mathbf{k}} + \mu)/w)$ is the order parameter.

A BCS-like equation for $\Delta_{\mathbf{k}} = \Delta(\xi_{\mathbf{k}})$ has the half polaron bandwidth w as a cutting parameter in the integral, rather than the Debye temperature:

$$\Delta(\xi) = \int_{-w-\mu}^{w-\mu} \mathrm{d}\eta N_{\mathrm{p}}(\eta) V(\xi, \eta) \frac{\Delta(\eta)}{2\sqrt{\eta^2 + \Delta(\eta)^2}} \tanh \frac{\sqrt{\eta^2 + \Delta^2(\eta)}}{2T}, \tag{5.22}$$

where $V(\xi, \eta) = -v_0 - zv_1(\xi + \mu)(\eta + \mu)/w^2$.

The critical temperature T_{c} is found as [130]

$$T_{\mathrm{c}} \approx 1.14w\sqrt{1 - \frac{\mu^2}{w^2}} \exp\left(\frac{2w}{v_0 + zv_1\mu^2/w^2}\right), \tag{5.23}$$

if $v_0 + zv_1\mu^2/w^2 < 0$. Hence, the polaronic superconductivity exists even in the case of the on-site repulsion, $v_0 > 0$, if this repulsion is less than the total intersite attraction, $z|v_1|$. There is a nontrivial dependence of T_{c} on doping. With a constant density of states in the polaron band, the Fermi energy $\epsilon_{\mathrm{F}} \approx \mu$ is expressed via the number of polarons per atom n_{p} as

$$\mu = w(n_{\mathrm{p}} - 1), \tag{5.24}$$

so that

$$T_{\mathrm{c}} \simeq 1.14w\sqrt{n_{\mathrm{p}}(2 - n_{\mathrm{p}})} \exp\left(\frac{2w}{v_0 + zv_1[n_{\mathrm{p}} - 1]^2}\right). \tag{5.25}$$

It has two maxima as a function of n_{p} separated by a deep minimum in the half-filled band ($n_{\mathrm{p}} = 1$), where the nearest-neighbor contributions to pairing are canceled.

An important feature of T_{c} in polaronic and bipolaronic superconductors is the unusual isotope effect [196]. With increasing ion mass, the (bi)polaron mass increases and the Bose–Einstein condensation temperature $T_{\mathrm{c}} \propto 1/m^*$ decreases in the bipolaronic superconductor (Sect. 5.5). On the contrary, in a polaronic superconductor an increase of the ion mass leads to a band narrowing enhancing the polaron density of states and *increasing* T_{c} (5.25). Hence, the isotope exponent of T_{c}, $\alpha = -\mathrm{d}\ln T_{\mathrm{c}}/\mathrm{d}\ln M$ can distinguish the BCS-like polaronic superconductivity with $\alpha < 0$, and the Bose–Einstein condensation of small bipolarons with $\alpha > 0$. Moreover, α could be larger than the BCS value $\alpha > 0.5$, in bipolaronic superconductors. Using (5.25), one can link the isotope effect on T_{c} with the isotope effect on the carrier mass, α_{m^*} (see Sect. 3.5 and (3.72)).

Another important conclusion is that the highest superconducting transition temperature is attained in polaronic superconductors at the crossover between the BCS weak-coupling regime and the bipolaronic (strong-coupling) superconductivity [130].

5.4 Lattice Bipolarons and Competing Orders

In models with strong attractions, there is a possibility of clustering. A finite-range EPI combined with the Coulomb repulsion can cause clustering of polarons into finite-size strings [351,352] and other mesoscopic textures [214]. Formation of polaronic clusters can be analytically studied in the strong-coupling regime in the framework of a generic Fröhlich–Coulomb model (FCM) (Sect. 4.4) on a two-dimensional lattice of ideal octahedra (Fig. 5.4) that can be regarded as a simplified model of the copper–oxygen perovskite layer [311].

The FCM Hamiltonian including the coupling with different vibration modes α is written as

$$
\begin{aligned}
H = &- \sum_{\mathbf{n}\neq\mathbf{n}'} \left[T(\mathbf{n}-\mathbf{n}')c_{\mathbf{n}}^{\dagger}c_{\mathbf{n}'} - \frac{1}{2}V_{\mathrm{c}}(\mathbf{n}-\mathbf{n}')c_{\mathbf{n}}^{\dagger}c_{\mathbf{n}}c_{\mathbf{n}'}^{\dagger}c_{\mathbf{n}'} \right] \\
&- \sum_{\alpha,\mathbf{nm}} \omega_{\alpha} g_{\alpha}(\mathbf{m}-\mathbf{n})(\mathbf{e}_{\alpha}\cdot\mathbf{u}_{\mathbf{m}-\mathbf{n}})c_{\mathbf{n}}^{\dagger}c_{\mathbf{n}}(d_{\mathbf{m}\alpha}^{\dagger}+d_{\mathbf{m}\alpha}) \\
&+ \sum_{\mathbf{m}\alpha} \omega_{\alpha}\left(d_{\mathbf{m}\alpha}^{\dagger}d_{\mathbf{m}\alpha}+1/2\right),
\end{aligned}
\tag{5.26}
$$

where $\mathbf{e}_{\alpha}$ is the polarization vector of the αth vibration coordinate.

Due to poor screening, the hole–ion interaction can be taken as purely coulombic:

$$
g_{\alpha}(\mathbf{m}-\mathbf{n}) = \frac{\kappa_{\alpha}}{|\mathbf{m}-\mathbf{n}|^2},
$$

where $\alpha = x, y, z$ with $\kappa_x = \kappa_y = \kappa_z/\sqrt{2}$ accounting for the experimental fact that c-axis (z-polarized) phonons couple to in-plane holes stronger than

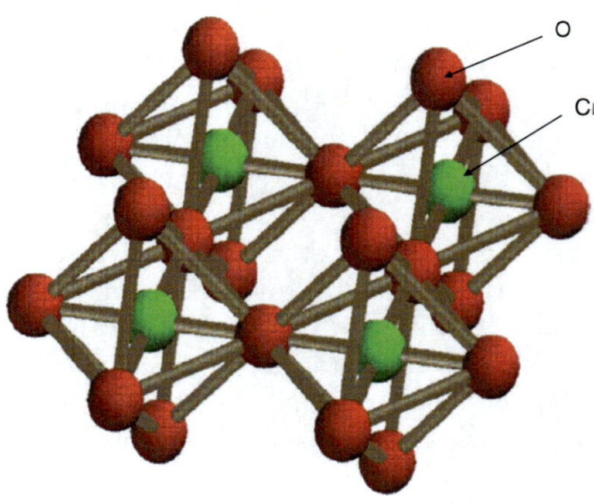

Fig. 5.4. Simplified model of the copper–oxygen perovskite layer (reproduced from [311], © IoP, 2002)

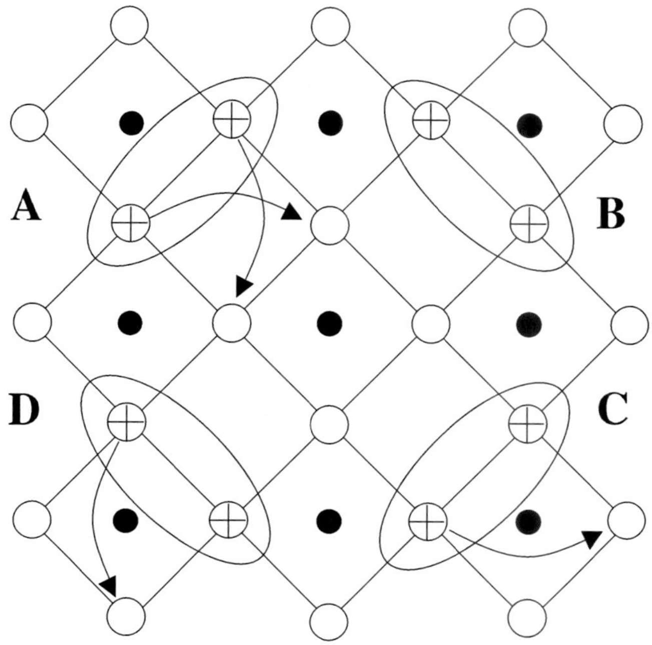

Fig. 5.5. Four degenerate in-plane bipolaron configurations A, B, C, and D. Some single-polaron hoppings are indicated by *arrows* (reproduced from [311], © IoP, 2002)

others. The direct hole–hole repulsion is

$$V_c(\mathbf{n}-\mathbf{n}') = \frac{V_c}{\sqrt{2}|\mathbf{n}-\mathbf{n}'|},$$

so that the repulsion between two holes in the nearest-neighbor (NN) configuration is V_c. The nearest-neighbor hopping T_{NN}, the next-nearest-neighbor (NNN) hopping across copper T_{NNN}, and the hopping between the pyramids T'_{NNN} have been included (Fig. 5.5).

The polaron level shift in this model is given by the lattice sum:

$$E_{\text{p}} = 2\kappa_x^2\omega_0 \sum_{\mathbf{m}} \left(\frac{1}{|\mathbf{m}-\mathbf{n}|^4} + \frac{h^2}{|\mathbf{m}-\mathbf{n}|^6} \right) = 31.15\kappa_x^2\omega_0, \tag{5.27}$$

where the factor 2 accounts for two layers of apical sites, and the in-plane lattice constant is $a = 1$ and $\omega_\alpha = \omega_0$. For reference, the Cartesian coordinates are $\mathbf{n} = (n_x + 1/2, n_y + 1/2, 0)$, $\mathbf{m} = (m_x, m_y, h)$, and n_x, n_y, m_x, m_y are integers. The polaron–polaron attraction is

$$V_{\text{ph}}(\mathbf{n}-\mathbf{n}') = 4\omega_0\kappa_x^2 \sum_{\mathbf{m}} \frac{h^2 + (\mathbf{m}-\mathbf{n}')\cdot(\mathbf{m}-\mathbf{n})}{|\mathbf{m}-\mathbf{n}'|^3|\mathbf{m}-\mathbf{n}|^3}. \tag{5.28}$$

Performing the lattice summations for the NN, NNN, and NNN$'$ configurations, one finds $V_{\mathrm{ph}} = 1.23E_{\mathrm{p}}$, $0.80E_{\mathrm{p}}$, and $0.82E_{\mathrm{p}}$, respectively. As a result, we obtain a net interpolaron interaction as $v_{\mathrm{NN}} = V_{\mathrm{c}} - 1.23E_{\mathrm{p}}$, $v_{\mathrm{NNN}} = (V_{\mathrm{c}}/\sqrt{2}) - 0.80E_{\mathrm{p}}$, $v'_{\mathrm{NNN}} = (V_{\mathrm{c}}/\sqrt{2}) - 0.82E_{\mathrm{p}}$, and the mass renormalization exponents (see below) as $g^2_{\mathrm{NN}} = 0.38(E_{\mathrm{p}}/\omega)$, $g^2_{\mathrm{NNN}} = 0.60(E_{\mathrm{p}}/\omega)$, and $(g'_{\mathrm{NNN}})^2 = 0.59(E_{\mathrm{p}}/\omega)$. At $V_{\mathrm{c}} > 1.23E_{\mathrm{p}}$, no bipolarons are formed and the system is a polaronic Fermi liquid. Polarons tunnel in the *square* lattice with the renormalized hopping integrals $t = T_{\mathrm{NN}} \exp(-0.38E_{\mathrm{p}}/\omega)$ and $t' = T_{\mathrm{NNN}} \exp(-0.60E_{\mathrm{p}}/\omega)$ for NN and NNN hoppings, respectively. The polaron mass is $m^* \propto 1/(t+2t')$.

If $V_{\mathrm{c}} < 1.23E_{\mathrm{p}}$, then intersite NN bipolarons form. The intersite bipolarons tunnel in the plane via four resonating (degenerate) configurations A, B, C, and D, as shown in Fig. 5.5. In the first order of the renormalized hopping integral, one should retain only these lowest energy configurations and discard all the processes that involve configurations with higher energies. These intersite bipolarons already move in the *first* order of the single polaron hopping. This remarkable property is entirely due to the strong on-site repulsion and long-range electron–phonon interactions that leads to a nontrivial connectivity of the lattice. This fact combines with a weak renormalization of t' yielding a *superlight* bipolaron with the mass $m^{**} \propto \exp(0.60E_{\mathrm{p}}/\omega)$. We recall that in the Holstein model $m^{**} \propto \exp(2E_{\mathrm{p}}/\omega)$. Thus, the mass of the small Fröhlich bipolaron in the perovskite layer scales approximately as a *cubic root* of that of the Holstein bipolaron.

At even stronger EPI, $V_{\mathrm{c}} < 1.16E_{\mathrm{p}}$, NNN bipolarons become stable. More importantly, holes can now form three- and four-particle clusters. The dominance of the potential energy over kinetic in the transformed Hamiltonian enables us to readily investigate these many-polaron cases. Three holes placed within one oxygen square have four degenerate states with the energy $2(V_{\mathrm{c}}-1.23E_{\mathrm{p}})+V_{\mathrm{c}}/\sqrt{2}-0.80E_{\mathrm{p}}$. The first-order polaron hopping processes mix the states resulting in a ground-state linear combination with the energy $E_3 = 2.71V_{\mathrm{c}}-3.26E_{\mathrm{p}}-\sqrt{4t^2+t'^2}$. It is essential that between the squares such triads could move only in higher orders of polaron hopping. In the first order, they are immobile. A cluster of four holes has only one state within a square of oxygen atoms. Its energy is $E_4 = 4(V_{\mathrm{c}}-1.23E_{\mathrm{p}})+2(V_{\mathrm{c}}/\sqrt{2}-0.80E_{\mathrm{p}}) = 5.41V_{\mathrm{c}}-6.52E_{\mathrm{p}}$. This cluster, as well as all bigger ones, are also immobile in the first order of polaron hopping. Hence, a strong EPI combined with the Coulomb repulsion could cause clustering of polarons into finite-size mesoscopic textures. Importantly at distances much larger than the lattice constant, the polaron–polaron interaction is always repulsive, and the formation of infinite clusters, stripes, or strings is prohibited [351].

There is strong experimental evidence for superlight intersite bipolarons in cuprate superconductors (Sect. 6.1), where they form in-plane oxygen–apex oxygen pairs (so-called apex bipolarons [59, 353] and in-plane oxygen–oxygen pairs [25, 311]). Analytical [311, 351] and QMC [214] studies with realistic lattice deformations and the Coulomb repulsion show that real-space pairs dominate over entire phase diagram since bipolarons effectively repel each other (4.18) (for more details on polaronic textures, see [277, 354, 355]). Nevertheless, there is evidence in favor of small-polaron lattices in some lattice structures (e.g., nickelates) with low-dimensional energy spectrum and significant densities of small polarons [354, 356, 357]. Long-range ordering of polarons is especially important in 1D systems. In this respect, 1D Holstein model of spinless fermions and 1D HHM have been investigated in much detail using the density matrix renormalization group [358, 359], kernel polynomial and cluster approaches [360], Lanczos diagonalization [361], and QMC [362] with a focus on the competition of conducting, insulating, and (bi)polaron charge-ordered phases. This issue was also discussed in infinite dimension using DMFT [206, 363].

5.5 Bipolaronic Superconductivity

In the strong-coupling regime, where the BCS electron–phonon coupling constant is relatively large, $\lambda \gtrsim 1$, pairing is individual [9], in contrast with the collective Cooper pairing [342]. Bipolarons survive even in the normal state above their Bose–Einstein condensation temperature representing a simplest "cluster" of carriers. These and bigger clusters can be localized by disorder below the mobility edge. However, bipolarons propagate as the Bloch states above the mobility edge.

In the subspace with no single polarons, the Hamiltonian of electrons strongly coupled with phonons is reduced to the bipolaronic Hamiltonian (4.18) written in terms of creation and annihilation bipolaron operators [271]. There could be additional spin quantum numbers $S = 0, 1$; $S_z = 0, \pm 1$, which should be added to the definition of the bipolaron operator $b_{\mathbf{m}}$ in the case of intersite bipolarons. Bipolarons are not perfect bosons since their operators commute as

$$b_{\mathbf{m}} b_{\mathbf{m}}^{\dagger} + b_{\mathbf{m}}^{\dagger} b_{\mathbf{m}} = 1, \tag{5.29}$$

$$b_{\mathbf{m}} b_{\mathbf{m}'}^{\dagger} - b_{\mathbf{m}'}^{\dagger} b_{\mathbf{m}} = 0 \tag{5.30}$$

for $\mathbf{m} \neq \mathbf{m}'$. This makes useful the pseudospin analogy [271], $b_{\mathbf{m}}^{\dagger} = S_{\mathbf{m}}^{x} - \mathrm{i}S_{\mathbf{m}}^{y}$ and $b_{\mathbf{m}}^{\dagger} b_{\mathbf{m}} = (1/2) - S_{\mathbf{m}}^{z}$ using the pseudospin 1/2 operators $S^{x,y,z} = (1/2)\tau_{1,2,3}$(here τ_i are Pauli matrices). $S_{\mathbf{m}}^{z} = 1/2$ corresponds to an empty site $\mathbf{m}$ and $S_{\mathbf{m}}^{z} = -1/2$ to a site occupied by the bipolaron. Spin operators preserve the bosonic nature of bipolaron operators, when bipolarons are on different sites, and their fermionic internal structure. Replacing bipolaron operators by spin operators, one transforms the bipolaronic Hamiltonian into the familiar anisotropic Heisenberg Hamiltonian:

$$H_{\mathrm{b}} = \sum_{\mathbf{m} \neq \mathbf{m}'} \left[\frac{1}{2} \bar{v}(\mathbf{m} - \mathbf{m}') S_{\mathbf{m}}^{z} S_{\mathbf{m}'}^{z} + t_{\mathrm{b}}(\mathbf{m} - \mathbf{m}')(S_{\mathbf{m}}^{x} S_{\mathbf{m}'}^{x} + S_{\mathbf{m}}^{y} S_{\mathbf{m}'}^{y}) \right], \tag{5.31}$$

where the position of the middle of the bipolaron band is taken as zero. This Hamiltonian has been investigated in detail as a relevant form for magnetism and also for quantum solids like a lattice model of ^{4}He. However, while in those cases the magnetic field is an independent thermodynamic variable, in our case the total "magnetization" is fixed, $N^{-1}\sum_{\mathbf{m}}\langle\langle S^z_{\mathbf{m}}\rangle\rangle = 1/2 - n_{\mathrm{b}}$, if the bipolaron density n_{b} is conserved.

Spin 1/2 Heisenberg Hamiltonian cannot be solved analytically. Complicated commutation rules for bipolaron operators make the problem hard, but not in the limit of low atomic density of bipolarons, $n_{\mathrm{b}} \ll 1$. In this limit, we can reduce the problem to a charged Bose gas (CBG) on a lattice [364] by transforming the bipolaronic Hamiltonian to a representation containing only the Bose operators $a_{\mathbf{m}}$ and $a^\dagger_{\mathbf{m}}$ defined as

$$b_{\mathbf{m}} = \sum_{k=0}^{\infty} \beta_k (a^\dagger_{\mathbf{m}})^k a^{k+1}_{\mathbf{m}}, \tag{5.32}$$

where $a_{\mathbf{m}}a^\dagger_{\mathbf{m}'} - a^\dagger_{\mathbf{m}'}a_{\mathbf{m}} = \delta_{\mathbf{m},\mathbf{m}'}$. The first few coefficients β_k are found as $\beta_0 = 1$, $\beta_1 = -1$, $\beta_2 = 1/2 + \sqrt{3}/6$. One can also introduce bipolaron and boson Ψ-operators as $\Phi(\mathbf{r}) = N^{-1/2}\sum_{\mathbf{m}}\delta(\mathbf{r}-\mathbf{m})b_{\mathbf{m}}$ and $\Psi(\mathbf{r}) = N^{-1/2}\sum_{\mathbf{m}}\delta(\mathbf{r}-\mathbf{m})a_{\mathbf{m}}$, and write down the bipolaronic Hamiltonian as

$$H_{\mathrm{b}} = \int \mathrm{d}\mathbf{r} \int \mathrm{d}\mathbf{r}' \Psi^\dagger(\mathbf{r}) t(\mathbf{r}-\mathbf{r}') \Psi(\mathbf{r}') + H_{\mathrm{d}} + H_{\mathrm{k}} + H^{(3)}, \tag{5.33}$$

where

$$H_{\mathrm{d}} = \frac{1}{2}\int \mathrm{d}\mathbf{r} \int \mathrm{d}\mathbf{r}' \bar{v}(\mathbf{r}-\mathbf{r}')\Psi^\dagger(\mathbf{r})\Psi^\dagger(\mathbf{r}')\Psi(\mathbf{r}')\Psi(\mathbf{r}) \tag{5.34}$$

is the dynamic part,

$$\begin{aligned} H_{\mathrm{k}} = \frac{2}{N}\int \mathrm{d}\mathbf{r} \int \mathrm{d}\mathbf{r}' t_{\mathrm{b}}(\mathbf{r}-\mathbf{r}') \\ \times \left[\Psi^\dagger(\mathbf{r})\Psi^\dagger(\mathbf{r}')\Psi(\mathbf{r}')\Psi(\mathbf{r}') + \Psi^\dagger(\mathbf{r})\Psi^\dagger(\mathbf{r})\Psi(\mathbf{r})\Psi(\mathbf{r}')\right] \end{aligned} \tag{5.35}$$

is the kinematic (hard-core) part due to the "imperfect" commutation rules and $H^{(3)}$ includes three- and higher-body collisions. Here

$$t_{\mathrm{b}}(\mathbf{r}-\mathbf{r}') = \sum_{\mathbf{k}} \epsilon_{\mathrm{b}}(\mathbf{k}) \mathrm{e}^{\mathrm{i}\mathbf{k}\cdot(\mathbf{r}-\mathbf{r}')},$$

$$\bar{v}(\mathbf{r}-\mathbf{r}') = \frac{1}{N}\sum_{\mathbf{k}} \bar{v}_{\mathbf{k}} \mathrm{e}^{\mathrm{i}\mathbf{k}\cdot(\mathbf{r}-\mathbf{r}')},$$

$\bar{v}_{\mathbf{k}} = \sum_{\mathbf{m}\neq 0}\bar{v}(\mathbf{m})\exp(\mathrm{i}\mathbf{k}\cdot\mathbf{m})$ is the Fourier component of the dynamic interaction, and

$$\epsilon_{\mathrm{b}}(\mathbf{k}) = \sum_{\mathbf{m}\neq 0} t_{\mathrm{b}}(\mathbf{m})\exp(-\mathrm{i}\mathbf{k}\cdot\mathbf{m}) \tag{5.36}$$

is the bipolaron band dispersion. $H^{(3)}$ contains more than four field operators. In the dilute limit, $n_\mathrm{b} \ll 1$, only two-particle interactions are essential which include the short-range kinematic and direct density–density repulsions. Since $\bar{v}$ already contains the short-range part $v^{(2)}$ (4.18), the kinematic contribution can be included in the definition of $\bar{v}$. As a result, H_b is reduced to the Hamiltonian of interacting hard-core charged bosons tunneling in the narrow band.

To describe electrodynamics of bipolarons, one introduces the vector potential $\mathbf{A}(\mathbf{r})$ using the Peierls substitution [237]:

$$t_\mathrm{b}(\mathbf{m}-\mathbf{m}') \rightarrow t_\mathrm{b}(\mathbf{m}-\mathbf{m}')\mathrm{e}^{\mathrm{i}2e\mathbf{A}(\mathbf{m})\cdot(\mathbf{m}-\mathbf{m}')},$$

which yields

$$t_\mathrm{b}(\mathbf{r}-\mathbf{r}') \rightarrow t_\mathrm{b}(\mathbf{r},\mathbf{r}') = \sum_{\mathbf{k}} \epsilon_\mathrm{b}(\mathbf{k}-2e\mathbf{A})\mathrm{e}^{\mathrm{i}\mathbf{k}\cdot(\mathbf{r}-\mathbf{r}')} \tag{5.37}$$

in the real space. If the magnetic field is weak, we can expand $\epsilon_\mathrm{b}(\mathbf{k})$ in the vicinity of $\mathbf{k} = 0$ to obtain $t_\mathrm{b}(\mathbf{r},\mathbf{r}') \approx -\left[\nabla + 2\mathrm{i}e\mathbf{A}(\mathbf{r})\right]^2 \delta(\mathbf{r}-\mathbf{r}')/m^{**}$, where $1/m^{**} = \left(\mathrm{d}^2\epsilon_\mathrm{b}(\mathbf{k})/\mathrm{d}k^2\right)_{k\rightarrow 0}$ is the inverse bipolaron mass. Assuming a parabolic dispersion near the bottom of the band, $\epsilon_\mathrm{b}(\mathbf{k}) \sim k^2$, yields

$$\begin{aligned} H_\mathrm{b} \approx &- \int \mathrm{d}\mathbf{r}\Psi^\dagger(\mathbf{r})\left\{\frac{[\nabla+2\mathrm{i}e\mathbf{A}(\mathbf{r})]^2}{2m^{**}} + \mu\right\}\Psi(\mathbf{r}) \\ &+ \frac{1}{2}\int \mathrm{d}\mathbf{r}\mathrm{d}\mathbf{r}'\bar{v}(\mathbf{r}-\mathbf{r}')\Psi^\dagger(\mathbf{r})\Psi^\dagger(\mathbf{r}')\Psi(\mathbf{r})\Psi(\mathbf{r}'), \end{aligned} \tag{5.38}$$

where we add the *bipolaronic* chemical potential μ. The bipolaron–bipolaron interaction is the Coulomb repulsion, $\bar{v}(\mathbf{r}) \sim 1/(\epsilon_0 r)$ at large distances, and the hard-core effects are not important in the dilute limit. Hence, H_b describes the CBG with the effective boson mass m^{**} and charge $2e$, which exhibits the equilibrium features of a superconductor [365]. Moreover, this gas is almost ideal since $\epsilon_0 \gg 1$ in ionic lattices like cuprates, so that the Coulomb repulsion of mobile bipolarons is weak compared with their kinetic energy.

At relatively high densities, the bipolaronic hard-core Bose gas can transform into a charge-density wave supersolid [271], similar to a supersolid proposed for ^{4}He [366], with a growing tendency to a phase separation with doping (different phase diagrams of hard-core bipolarons or bosons on a lattice can be found in [277, 367–370], and references therein).

5.6 Polarons and Spin Effects

An important feature of contemporary materials like cuprates and manganites is the coupling of the carrier charge degrees of freedom to a dense background of magnetically correlated spins (ferro in colossal magnetoresistance manganites and antiferro in cuprates). Evidence for the resulting spin correlations and underlying Hund's rule and/or Mott–Hubbard strong local Coulomb interactions is overwhelming in these materials (see, e.g., [25, 218, 293, 371–376]).

Since EPI proved to be also very strong (see [11, 59, 197–199, 201, 211, 216, 218, 219, 377–390], and references therein), its interplay with electron exchange correlations in the formation of dressed quasiparticles in manganites, cuprates, and many other materials, including conducting polymers and ferroelectrics, is becoming the focus of great attention (for recent reviews, see [35, 58, 391]). The coupling of lattice polarons to the spin degrees of freedom alters their tunneling and pairing characteristics. A lattice disorder introduces additional complexity to the polaron problem because interference of impurity potential with lattice distortion, which accompanies the polaron movement, can contribute to the polaron and bipolaron localization on impurity [26, 28, 392–395]. Also networks of local distortions correlated by elastic fields in doped transition metal oxides [396] and self-organized discrete dopant networks [397] lead to multiscale complexity for key materials.

The foregoing points have been the central theme and main emphasis of a large body of theoretical works on polaron physics in strongly correlated electron models that are potentially relevant to the manganites and cuprates. Apart from the strong-coupling bipolaronic limit discussed above, the interplay of EPI with the on-site Coulomb repulsion in the intermediate region of parameters has been studied in the framework of the Holstein–Hubbard (HHM) and the Holstein t-J (HtJM) models. In the latter model, the strong electron–electron correlations are described by

$$H_{\mathrm{tJ}} = -t \sum_{\langle \mathbf{m},\mathbf{n}\rangle ,s} \left(\tilde{c}^{\dagger}_{\mathbf{m},s}\tilde{c}_{\mathbf{n},s} + \mathrm{H.c.}\right) + J \sum_{\langle \mathbf{m},\mathbf{n}\rangle} \left(\vec{S}_{\mathbf{m}}\vec{S}_{\mathbf{n}} - \frac{1}{4}\tilde{n}_{\mathbf{m}}\tilde{n}_{\mathbf{n}}\right), \qquad (5.39)$$

acting in a projected Hilbert space, i.e., $\tilde{c}_i^{(\dagger)} = c^{(\dagger)}_{\mathbf{m},s}(1-\tilde{n}_{\mathbf{m},-s})$, $\tilde{n}_{\mathbf{m}} = \sum_s \tilde{c}_i^{\dagger}\tilde{c}_i$, and $\vec{S}_i = \sum_{s,s'} \tilde{c}^{\dagger}_{\mathbf{m},s}\vec{\tau}_{ss'}\tilde{c}_{\mathbf{m},s'}$. Within the t-J model, the bare transfer amplitude of electrons (t) sets the energy scale for incoherent transport, while the Heisenberg interaction (J) allows for spin flips leading to coherent hole motion at the bottom of a band with an effective bandwidth determined by $J \ll t$.

To study polaronic effects in systems exhibiting besides strong antiferromagnetic exchange a substantial EPI, the Hamiltonian (5.39) is supplemented by a Holstein-type interaction term:

$$H = H_{\mathrm{tJ}} + \omega_0 \sum_{\mathbf{m}} [g(d^{\dagger}_{\mathbf{m}} + d_{\mathbf{m}})\,\tilde{h}_{\mathbf{m}} + (d^{\dagger}_{\mathbf{m}}d_{\mathbf{m}} + 1/2)] \qquad (5.40)$$

($\tilde{h}_{\mathbf{m}} = 1 - \tilde{n}_{\mathbf{m}}$ denotes the local density operator of the spinless hole). The resulting HtJM (5.40) takes the coupling to the hole as dominant source of the particle–lattice interaction. In the cuprate context an unoccupied site, i.e., a hole, corresponds to a Zhang–Rice singlet [398] formed by Cu $3d_{x^2-y^2}$ and O $2p_{x,y}$ hole orbitals.

In the framework of HHM, Zhong and Schuttler [399] demonstrated for the nearly half-filled band that the presence of antiferromagnetic spin correlations can cause polaronic carrier self-localization and strongly anharmonic lattice potential, already at moderate electron–phonon coupling strengths $\lambda \approx 0.2$–0.4. The anharmonic effects were analyzed in the framework of the Holstein and HHM models through a strong-coupling expansion up to fourth order in the hopping integrals [400]. Anharmonicity favors superconductivity relative to charge-density-wave order. Fehske *et al.* [401] showed that the Coulomb repulsion shifts the self-trapping transition to lower EPI strength, and the effective polaron mass below the transition is enhanced.

The formation of hole pairs in the planar t-J model was studied in the presence of in-plane (breathing modes) and out-of-plane (buckling modes) displacements [402], where strong evidence in favor of a stabilization of the two-hole bound pair by out-of-plane vibrations of the in-plane oxygens was found. On the contrary, the breathing modes weaken the binding energy of the hole pair. The tunneling dynamics of dopant-induced hole polarons that are self-localized by electron–phonon coupling in a two-dimensional antiferromagnet was studied by Yonemitsu *et al.* [403] using HtJM. Antiferromagnetic spin correlations in the original many-electron Hamiltonian are reflected by an attractive contribution to the first-neighbor charge interaction and by Berry phase factors that determine the signs of effective polaron tunneling matrix elements with implications for the doping-dependent isotope effect, pseudogap, and T_c of a superconducting polaron-pair condensate of the cuprate superconductors. More recently, self-localization of holes in HtJM was studied using ED and DQMC [35,58,404,405]. It was shown that the critical electron–phonon coupling decreases with increasing J. The spin correlation functions in the localized region can be understood within a percolation picture where antiferromagnetic order persists up to a substantial hole doping [405]. EPI can also affect phase separations in the form of tiles and stripe phases inherent for the t-J model. The stripes are stabilized by half-breathing modes, while the tiles arise due to the development of extended breathing modes [406].

EPI itself and phonons are also affected by the electron–electron correlations. In the weak- and intermediate-correlation regimes, one finds that the on-site Coulomb interaction U acts to effectively suppress EPI due to screening [292, 407]. However, entering the strong-correlation regime, EPI stops decreasing and begins to substantially increase as a function of U, leading to an effective EPI which is peaked in the forward direction [407]. The phonon propagator shows a characteristic softening near the metal to bipolaronic transition but there is very little softening on the approach to the Mott transition [408]. Importantly, the interaction with c-axis polarized optical phonons in cuprates remains virtually unscreened since a poor mobility of carriers perpendicular to the CuO_2 planes, so that the Fröhlich EPI remains strong at any doping [59].

Other types of EPIs, in particular SSH interaction was also studied in strongly correlated systems [409]. The phase diagram of a Hamiltonian both with on-site repulsion between electrons and with hopping probabilities which depend on dynamically varying separations of neighboring lattice sites was obtained as a function of density, electron–electron, and electron–phonon interaction strengths. The superconducting regime is characterized by a pattern of short and long bonds correlated with resonating pair hopping.

The foregoing discussion is by no means complete. The theoretical works on strongly correlated EPI systems are one of the most hot topics in condensed matter physics, and many of them are currently in progress. Generally speaking, the changes of the quasiparticle properties due to the combined effects of hole–phonon/(pseudo)magnon correlations are expected to be very complex and have to be addressed using numerical techniques such as ED on finite lattices with the phonon truncated Hilbert space, QMC and DQMC. Dressed hole quasiparticles will show the characteristics of both "lattice" and "magnetic" (spin) polarons (for recent comprehensive reviews, see [35, 58]).

6

Polarons and Bipolarons in Advanced Materials

EPIs have been shown to be relevant in cuprate and other high-temperature superconductors through, e.g., isotope substitution experiments [197–199], high-resolution angle-resolved photoemission (ARPES) [201,382,383], a number of earlier optical [378,384,389], neutron scattering [388] and more recent inelastic scattering [385], and some other measurements [410]. In colossal magnetoresistance (CMR) manganites, isotope substitutions [390], X-ray and neutron scattering spectroscopies [379, 380], and a number of other experiments also show a significant effect of EPI on the physical properties (for review, see [411]). Therefore, it has been suggested that the long-range [13] and/or the molecular-type (e.g., [210]) EPIs play significant role in high-temperature superconductors (see [11, 59, 211, 216], and references therein), and in CMR manganites (see [218, 219, 377], and references therein).

6.1 Polarons in High-Temperature Superconductors

There are still many theories that attempt to explain the phenomenon of high-temperature superconductivity in cuprates and other related materials. In general, the pairing mechanism of carriers could be not only "phononic" as in the BCS theory [342] (left-hand upper corner in Fig. 6.1) or its strong-coupling bipolaronic extension [130] (right-hand upper corner in Fig. 6.1), but also "excitonic" [412, 413], "plasmonic" [414, 415], "magnetic" [416, 417], "kinetic" [418], or purely repulsive Coulomb due to a mirror-nested Fermi surface [419]. The BCS theory like any mean-field theory is rather universal, so that it describes well the cooperative quantum phenomenon of superconductivity even with these nonphononic mechanisms, if the coupling is weak (left-hand lower corner in Fig. 6.1). The main motivation behind these concepts is that high superconducting critical temperature, T_{c}, could be achieved by replacing phonons in the conventional BCS theory by higher-frequency bosonic modes, such as plasmons, spin waves (pseudomagnons), or even by the direct Coulomb repulsion combined with unconventional pairing symmetries.

A.S. Alexandrov and J.T. Devreese, *Advances in Polaron Physics*, Springer Series in Solid-State Sciences 159,
DOI: 10.1007/978-3-642-01896-1_6, © Springer-Verlag Berlin Heidelberg 2010

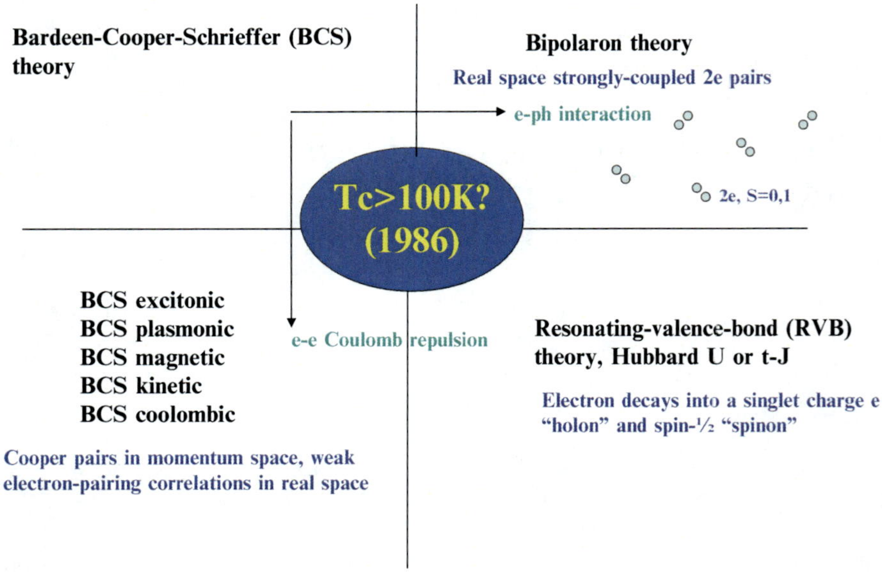

Fig. 6.1. A few theories of high-temperature superconductivity. The highest T_c is predicted in the BCS to bipolaron crossover [130] for the phonon pairing mechanism (*upper half* of the diagram). *Lower half* of the diagram represents a number of nonphononic mechanisms of pairing

Actually, following original proposal by P.W. Anderson [293], many authors [420] assumed that the electron–electron interaction in novel superconductors is very strong but repulsive and it provides high-T_c without any phonons (right-hand lower corner in Fig. 6.1). A motivation for this concept can be found in the earlier work by Kohn and Luttinger [421], who showed that the Cooper pairing of repulsive fermions is possible. However, the same work clearly showed that T_c of repulsive fermions is extremely low, well below the mK scale. Nevertheless, the BCS and BCS-like theories (including the Kohn–Luttinger consideration) heavily rely on the Fermi-liquid model of the *normal* state, which fails in many high-temperature superconductors. If the normal state is not the Fermi liquid, then there is no direct reason to reject the assumption. In fact, there is little doubt that strong on-site repulsive correlations (Hubbard U) are an essential feature of the cuprates. Indeed, all undoped cuprate compounds are *insulators* with the insulating gap about 2 eV or so. But if the repulsive correlations are weak, one would expect a metallic behavior of a half-filled d-band of copper in cuprates, or, at most, a much smaller gap caused by lattice and spin distortions (i.e., due to charge and/or spin density waves [422, 423]). It is a strong on-site repulsion of d-electrons in cuprates which results in their parent insulating "Mott" state. When on-site correlations are strong and dimensionality is low, there is an alternative to the usual Fermi liquid. In Anderson's resonating-valence-bond (RVB) model [293], the ground state supports "topological solitons" (the so-called spinons and holons), such as occur in one-dimensional Hubbard model. Theoretically, holons could be paired by a superexchange interaction without any additional glue like phonons or spin waves [424].

To discriminate one theory with respect to another, one has to rely on experimental facts and/or on exact theoretical results. Some variational Monte Carlo (VMC) simulations with a (projected) BCS-type trial wave function (see, e.g., [420] and references therein), and a number of other analytical and numerical studies appeared to back up superexchange pairing. However, recent studies by Aimi and Imada [425], using an advanced sign-problem-free Gaussian–Basis Monte Carlo (GBMC) algorithm, have shown that these variational methods, as well as other approximations, overestimated the normal state energy and therefore overestimated the condensation energy by at least an order of magnitude, so that the Hubbard model does not account for high-temperature superconductivity. The ground state of the model is a normal Fermi liquid with no superconductivity and no stripes. This remarkable result is in line with earlier numerical studies using the auxiliary-field quantum (AFQMC) [426] and constrained-path (CPMC) [427] Monte Carlo methods, none of which found superconductivity in the Hubbard model.

On the other hand, a growing number of observations point to the possibility that high-T_c cuprate superconductors [428] may not be conventional BCS superconductors [342]. Some of their features could be attributed to the Bose–Einstein condensation (BEC) of real-space mobile bipolarons [9,25,59,429]. A possible fundamental origin of such strong departure of the cuprates from conventional BCS behavior is the unscreened Fröhlich EPI, providing the polaron level shift E_p of the order of 1 eV [59,281,377], routinely neglected in the Hubbard U and t-J models [420]. This interaction with c-axis polarized optical phonons is virtually unscreened because the upper limit for the out-of-plane plasmon frequency ($\lesssim$200 cm^{-1} [430]) is well below the characteristic frequency of optical phonons, $\omega_0 \approx$ 400–1,000 cm^{-1}. Since screening is poor, the magnetic interaction remains small compared with the Fröhlich EPI at any doping of cuprates. Consequently, to build an adequate theory of high-temperature superconductivity, the long-range Coulomb repulsion and the *unscreened* EPI should be treated on an equal footing with the short-range Hubbard U. When these interactions are strong compared with the kinetic energy of carriers, this Coulomb–Fröhlich model predicts the ground state in the form of mobile intersite bipolarons [59,311] (see Sect. 4.4).

Most compelling evidence for (bi)polaronic carries in novel superconductors is the discovery of a substantial isotope effect on the carrier mass [197–199] predicted for the (bi)polaronic conductors in [196]. High-resolution ARPES [201, 382, 383] provides another piece of evidence for the strong EPI in cuprates [431] apparently with c-axis polarized optical phonons [383]. These as well as optical [378,384,389], neutron scattering [385,388], tunneling [432] spectroscopies of cuprates, and recent pump-probe experiments [410] unambiguously show that lattice vibrations play a significant though unconventional role in high-temperature superconductors. The interpretation of the optical spectra of high-T_c materials (measured in [338]) in the normal phase, and of manganites (measured in [339]) as due to many-polaron absorption [92] strengthens the view that the Fröhlich EPI is important in those structures. Operating together with a shorter-range deformation potential and molecular-type (e.g., Jahn–Teller [216]) EPIs, the Fröhlich EPI readily overcomes the Coulomb repulsion at a short distance about the lattice constant providing small but yet mobile bipolarons (Sect. 4.4).

A parameter-free estimate of the Fermi energy using the magnetic field penetration depth found a very low value, $\epsilon_F \lesssim 100\,\mathrm{meV}$ [433] clearly supporting the real-space (i.e., individual) pairing in cuprate superconductors. There is strong experimental evidence for a gap in the normal state electron density of states of cuprates [25], which is known as the pseudogap. Experimentally measured pseudogaps of many cuprates are about 50 meV or larger [434, 435]. If one accepts that the pseudogap is about half of the pair binding energy [286], then the condition for real-space pairing is well satisfied in most cuprates (typically $r_b \approx 0.2$–$0.4\,\mathrm{nm}$).

Also, magnetotransport and thermal magnetotransport data strongly support preformed bosons in cuprates. In particular, many high-magnetic-field studies revealed a non-BCS upward curvature of the upper critical field $H_{c2}(T)$ (see [436] for a review of experimental data), in accordance with the theoretical prediction for the Bose–Einstein condensation of charged bosons in the magnetic field [437]. The Lorenz number, $L = e^2\kappa_e/T\sigma$, differs significantly from the Sommerfeld value $L_e = \pi^2/3$ of the standard Fermi-liquid theory, if carriers are double-charged bosons [438]. Here, κ_e and σ are electron thermal and electrical conductivities, respectively. Alexandrov and Mott [438] predicted a rather low Lorenz number for bipolarons, $L \approx 0.15L_e$, due to the double elementary charge of bipolarons, and also due to their nearly classical distribution function above T_c. Direct measurements of the Lorenz number using the thermal Hall effect [439] produced the value of L just above T_c about the same as predicted by the bipolaron model, and its strong temperature dependence. This breakdown of the Wiedemann–Franz law is apparently caused by excited single polarons coexisting with bipolarons in the thermal equilibrium [440]. Also, unusual normal state diamagnetism uncovered by torque magnetometery [441, 442] has been convincingly explained as the normal state (Landau) diamagnetism of charged bosons [443].

Despite clear evidence for the existence of polarons in cuprate superconductors, no consensus currently exists concerning the microscopic mechanism of high-temperature superconductivity. As it has been emphasized in a number of early (1990s) and more recent studies of HHM and HtJM, the antiferromagnetic spin system promotes doping-induced polaron formation (Sect. 5.6). On the other hand, some works (e.g., [420]) suggest that EPI and polaron formation do not only not help, but hinder the pairing instability.

6.2 Polarons in Colossal Magnetoresistance Oxides

Ferromagnetic oxides, in particular manganese perovskites, have very large magnetoresistance near the ferromagnetic transition (for reviews, see [411, 444]). The effect observed in these materials was termed "colossal" magnetoresistance (CMR) to distinguish it from the giant magnetoresistance in metallic magnetic multilayers. The discovery raised expectations of a new generation of magnetic devices, and is a focus of extensive research aimed at describing the effect. Significant progress has been made in understanding the properties of CMR manganites, but many questions remain.

The ferromagnetic metal–insulator transition in manganites has long been thought as a consequence of the so-called double exchange mechanism (DEX), which results in a varying bandwidth of electrons in the Mn^{3+} d-shell as a function of temperature [445, 446]. Crystal fields split the Mn 3d orbitals into three localized t_{2g} orbitals, and two higher energy e_g orbitals which are hybridized with the oxygen p orbitals. Each manganese ion has a core spin of $S = 3/2$, and a fraction $(1 - x)$ (e.g., in $La_{1-x}Sr_xMnO_3$) have extra electrons in the e_g orbitals with spin parallel to the core spin due to the Hund's rule. The electron can hop to an adjacent Mn-site unoccupied e_g orbital with the probability that varies with the angle between the core spins.

However, it has been pointed out that DEX alone cannot account for CMR in $La_{1-x}Sr_xMnO_3$ [219], so that lattice polarons should be involved due to strong EPI arising from the Jahn–Teller effect [219, 220, 447, 448]. A tight-binding parametrization of LDA band theory combined with a dynamical mean-field treatment of correlations showed that manganites are characterized by a moderate Hund's coupling, and magnetic-order-driven changes in the kinetic energy may not be the cause of the observed "colossal" magnetoresistive and multiphase behavior [449]. The basic idea of [219] is that the electron–phonon coupling constant λ is large in the high-temperature paramagnetic state and the carriers are polarons, while the ferromagnetic order increases the bandwidth and thus decreases λ sufficiently for metallic behavior to occur below the Curie temperature T_C. Indeed, there is strong experimental evidence for exceptionally strong EPI in doped manganites from the isotope effects [390, 450], the Arrhenius behavior of the drift and Hall mobilities [451] in the paramagnetic phase above T_C, and many other experiments. Millis *et al.* [219, 448] and a great number of subsequent theoretical studies have combined DEX with the Jahn–Teller EPI in d-states arriving at the conclusion that the low-temperature ferromagnetic phase is a spin-polarized metal while the paramagnetic high-temperature phase is a polaronic insulator (for a review, see [218]).

A wide variety of experimental results and theoretical investigations have emphasized that competing physical interactions can cause the spontaneous emergence of electronic nanometer-scale structures in transition metal oxides like manganites [372]. Yunoki *et al.* [452], using a Kondo lattice Hamiltonian with ferromagnetic Hund's coupling as a model for manganites, identified a phase separation between hole-undoped antiferromagnetic and hole-rich ferromagnetic regions. Disorder on exchange and hopping amplitudes of the nondisordered strongly correlated system induces large coexisting metallic and insulating clusters with equal electronic density and percolative characteristics of manganites and related compounds [453]. Relevance of cooperative lattice effects and stress fields in phase separation was emphasized [396, 454].

Although it has been generally accepted that strong EPI plays an important role in manganites, a quantitative explanation of CMR has been lacking. The very nature of charge carriers in the ferromagnetic metallic state and in the paramagnetic state is still under intensive debate [218,450]. Some low-temperature optical [455,456], electron energy loss spectroscopy (EELS) [457], photoemission [458], and thermoelectric [459] measurements showed that the ferromagnetic phase of manganites is not a conventional metal. In particular, broad incoherent spectral features and a pseudogap were observed in ARPES. EELS [457], and O 1s X-ray absorption spectroscopy [460] consistently show that doped holes in manganites are of oxygen p character as expected for doped charge-transfer insulators, rather than d(Mn^{3+}) electrons. CMR has been observed in the ferromagnetic pyrochlore manganite $Tl_2Mn_2O_7$ [461], which has neither the mixed valence for DEX magnetic interaction nor the Jahn–Teller cations such as Mn^{3+}, raising a question of whether DEX combined with the Jahn–Teller EPI remains a relevant mechanism of CMR and ferromagnetism [377,462]. The existence of polarons has been demonstrated by atomic pair distribution [463], X-ray and neutron scattering studies [464–466] both above and well below T_C [467].

These and more recent experiments (e.g., the observation of that some samples of ferromagnetic manganites manifest an insulator-like optical conductivity even well below T_C [468]) further rule out DEX as the mechanism of CMR. The earlier of the above observations [390, 455–457, 460, 461] led to a novel theory of ferromagnetic/paramagnetic phase transition and CMR, based on the so-called current-carrier density collapse (CCDC) [377]. According to Alexandrov and Bratkovsky [377], one needs to consider the formation of small bipolarons in the paramagnetic state to explain CMR quantitatively. In the framework of CCDC p-holes are bound into heavy bipolarons above T_C due to the Fröhlich EPI. The resistivity peak near the transition and CMR are the result of the magnetic pair breaking below T_C, caused by the p–d spin-exchange interaction. Different from cuprates, oxygen-hole bipolarons are much heavier in manganites because the Fröhlich EPI is stronger [281]. They are readily localized by disorder, so it is mainly thermally excited single polarons that conduct in the paramagnetic phase. Upon temperature lowering single polarons polarize manganese spins at T_C via the exchange interaction J_{pd}, and the spin polarization of manganese breaks the bipolaronic singlets creating a spin-polarized *polaronic* metal.

CCDC quantitatively explained the temperature dependence of resistivity and CMR in the experimental range of external magnetic fields [377,469–471]. The oxygen isotope effect was observed in the low-temperature resistivity of

$La_{0.75}Ca_{0.25}$ MnO_3 and $Nd_{0.7}Sr_{0.3}MnO_3$ and explained with polaronic carriers in the ferromagnetic phase [200]. The current-carrier density collapse was observed in the Hall data in $La_{0.67}Ca_{0.33}MnO_3$ and $La_{0.67}Sr_{0.33}MnO_3$ [472], and the first-order phase transition at T_C, predicted by the theory [377], was established in the specific heat measurements [473]. Importantly, the character of the magnetic phase transition in $Tl_2Mn_2O_7$ pyrochlores has also been determined to be the first order [474]. CCDC also provided a simple explanation of the coexistence of high- and low-resistive phases as a mixture of the bipolaronic paramagnetic insulator and the polaronic ferromagnetic metal due to unavoidable disorder in doped manganites [469]. The concept of polaronic metal in ferromagnetic manganites [200, 377] has been substantiated by the angle-resolved photoemission spectroscopy data for the bilayer manganite $La_{1.2}Sr_{1.8}Mn_2O_7$, where a polaron metallic state below T_C has been clearly observed [475]. The observation of the pseudogap and nodal quasiparticles in ARPES of colossal magnetoresistive manganites [476], which have been considered as a characteristic feature of the copper oxide, further substantiates the analogy [477] between high-temperature superconducting and CMR oxides.

Nevertheless like in the case of the cuprates, there is still significant amount of work arguing that the major features of CMR manganite phenomenology can be reproduced by local spin degrees of freedom, coupled via Hund's rule interaction to a strongly correlated conduction electron system, where EPI plays only a supporting role [218, 478, 479].

6.3 Polarons in Nanostructures

6.3.1 Polaron Effects in Semiconductor Quantum Dots

Polarons in low-dimensional semiconductor structures and quasi-zero-dimensional systems like individual molecules have received significant attention in recent years. Quite generally, confinement enhances EPI and the tendency to polaron formation. Here, we touch upon just a few theoretical results on polaronic effects in nanostructures referring the reader to more comprehensive reviews and books [480–482].

A new aspect of the polaron concept has been investigated for semiconductor structures at nanoscale: the exciton–phonon states are not factorizable into an adiabatic product Ansatz, so that a *nonadiabatic* treatment is needed [483]. Considerable deviations of the oscillator strengths of the measured phonon-peak sidebands from the standard FC progression find a natural explanation within the nonadiabatic approach [483–485].

Experimental evidence of the enhanced phonon-assisted absorption due to effects of nonadiabaticity has been provided by the multiphonon photoluminescence (PL) spectra observed under selective excitation in self-assembled InAs/GaAs quantum dots [486] and by the photoluminescence excitation

(PLE) measurements on single self-assembled InAs/GaAs [487] and InGaAs/GaAs [488] quantum dots. The polaron concept was also invoked for the explanation of the PLE measurements on self-organized $In_xGa_{1-x}As/GaAs$ [489] and CdSe/ZnSe [490] quantum dots.

The stability of a strong-coupling singlet bipolaron was studied in two- and three-dimensional parabolic quantum dots using the Landau–Pekar variational method [491]. It was shown that the confining potential of the quantum dot reduces the stability of the bipolaron. A theory of bipolaron states in a spherical parabolic potential well was further developed applying the Feynman variational principle. The basic parameters of the bipolaron ground state (the binding energy, the number of phonons in the bipolaron cloud, and the bipolaron radius) were studied as a function of the radius of the potential well [492]. It was found that confinement can enhance the bipolaron binding energy, when the radius of a quantum dot is of the same order of magnitude as the polaron radius. A unified insight into the stability criterion for bipolaron formation in low-dimensionally confined media was provided by Senger and Ercelebi [493] using an adiabatic variational method for a pair of electrons immersed in a reservoir of bulk LO phonons and confined within an anisotropic parabolic potential box. Bipolaron formation in a two-dimensional lattice with harmonic confinement, representing a simplified model for a quantum dot, was investigated by means of QMC [494]. This method treats all interactions exactly and takes into account quantum lattice fluctuations. Calculations of the bipolaron binding energy reveal that confinement opposes bipolaron formation for weak electron–phonon coupling but abets a bound state at intermediate to strong coupling.

The ground-state energy and the optical conductivity spectra for a system with a finite number of interacting arbitrary-coupling large polarons in a spherical quantum dot were calculated using the path-integral formalism for identical particles [495–498]. Klimin *et al.* [499] have extended the memory-function approach to a system of arbitrary-coupling interacting polarons, confined to a parabolic potential. Importantly, the interplay of EPI and the Coulomb correlations within a confinement potential can lead to the clustering of polarons in multipolaron systems for strong EPI. The shell structure for a system of interacting polarons in a quantum dot is clearly revealed when analyzing the addition energy and the first-frequency moment of the optical conductivity.

Using the many-body path-integral formalism, the ground-state energy of an N-polaron system has been derived [500] at arbitrary-coupling strength in a confinement potential which is a combination of a parabolic potential and of a potential induced by a background charge. The many-body path-integral variational principle provides a rigorous upper bound for the ground-state energy of N polarons taking into account both the Fermi statistics and the Coulomb interaction between fermions. For finite N, the dependencies of the ground-state energy and of the polaron contribution to the ground-state energy on the average fermion density in a quantum dot are very similar to those for a polaron gas in bulk.

Finally, we mention the exciton–polaron formation in nanostructures and quantum-light sources studied recently by QMC in lattice models with short- or long-range carrier–phonon interaction (see [501] and references therein).

6.3.2 Correlated Polaron Transport Through Molecular Quantum Dots

It has been experimentally demonstrated that the low-bias conductance of molecules is dominated by resonant tunneling through coupled electronic and vibronic levels [502]. Conductance peaks caused by electron–vibron interactions have been seen in C_{60} [503]. Different electron–vibron interaction effects on the tunneling through molecular quantum dots (MQDs) have been studied (see, e.g., [504–509], and references therein). In the pioneering [506, 509] and subsequent studies of resonant-tunneling transmissions, EPI produced transmission sidebands peaks.

While correlation effects in transport through metallic quantum dots with repulsive electron–electron interactions received considerable attention in the past, and continue to be the focus of intense investigations, much less has been known about a role of attractive correlations between small polarons mediated by EPI in MQD. In the framework of the negative Hubbard U model [510], it has been found that the *attractive* electron correlations within the molecule could lead to a molecular *switching* effect where I–V characteristics have two branches with high and low current at the same bias voltage. The switching phenomenon has also been predicted by a theory of *correlated* polaron transport with a full account of both the Coulomb repulsion and EPI in MQD weakly coupled with electrodes [504, 505]. Alexandrov and Bratkovsky [504] have shown that while the phonon sidebands significantly modify the shape of hysteretic I–V curves in comparison with the negative-U Hubbard model, switching remains robust. It shows up at sufficiently low temperatures when the effective interaction of polarons in MQD is attractive and the molecular level is multiply degenerate. Importantly, the switching has not been found in *non-* and *twofold degenerate* MQDs either weakly [504] or strongly [511, 512] coupled with electrodes.

When the polaronic energy shift is very large, the effective charging energy of molecules can become negative, favoring ground states with even numbers of electrons. Koch *et al.* [513] have shown that charge transport through such molecules is dominated by tunneling of bipolarons which coexists with single-electron cotunneling. In asymmetric junctions, pair tunneling can be used for gate-controlled current rectification and switching. Electronic transport measurements of some single-molecule transistor devices [514] suggest that alternative switching mechanisms rather than (bi)polaronic effects could be involved in certain measurement configurations (see also [482, 515]).

7

Current Status of Polarons and Open Problems

At present, the basic properties of single polarons are well understood theoretically, and to a large extent they are analytically under control at all coupling. It is remarkable how the Fröhlich continuum polaron, one of the simplest examples of a *Quantum Field Theoretical* problem, as it basically consists of a single fermion interacting with a scalar Bose field, has resisted full analytical solution at all coupling since ~1950, when its Hamiltonian was first written. Although a mechanism for the optical absorption of Fröhlich polarons was already proposed a long time ago [73,98], some subtle characteristics were only clarified very recently [101] by combining numerical DQMC studies [100] and improved analytical methods [66,112] (Sect. 2.4.6). Of special interest are several sum rules derived for the optical conductivity spectra of arbitrary-coupling Fröhlich polarons [118, 119]. A variety of magneto-optical and transport experiments were successfully analyzed with Fröhlich polaron theory (see, e.g., [74, 83, 106, 516], and references therein).

The charge carriers in a rich variety of systems of reduced dimension and dimensionality (submicron- and nanostructures including heterojunctions, quantum wells, quantum wires, quantum dots, etc.) turn out to be Fröhlich polarons. Several scaling relations were derived [121], which connect polaron characteristics (the self-energy, the effective mass, the impedance, and the mobility) in different dimensions.

The Fröhlich polaron has led to many generalizations. The stability region of the Fröhlich large bipolaron is now firmly established [296, 301, 302] (Sect. 4.3). Here, the surprise is double (cf. [302, 310]) (a) only in a very limited sector of the phase diagram (Coulomb repulsion vs. α), the bipolaron is stable and (b) most traditional Fröhlich polaron materials (alkali halides and the like) lie completely outside (and "far" from) this bipolaron stability sector, but several high-T_c superconductors lie very close and even inside this very restricted area of the stability diagram. This should be a very hopeful sign for bipolaronic quasiparticles in the high-T_c superconductors.

The many-body theory for continuum polarons has been established for weak coupling and it became clear how – in this limit – this problem can be

A.S. Alexandrov and J.T. Devreese, *Advances in Polaron Physics*, Springer Series in Solid-State Sciences 159,
DOI: 10.1007/978-3-642-01896-1_7, © Springer-Verlag Berlin Heidelberg 2010

reduced to the study of the structure factor of a uniform electron gas [320]. In materials like high-T_c cuprates and manganites, a gas of interacting Fröhlich polarons has been invoked to quantitatively analyze optical experiments [92]. For stronger coupling, the problem remains highly cumbersome. The optical response of a many-polaron gas was analyzed in [341] using variational parameters of a single-polaron Feynman model. Progress has been made using path-integral approaches to the many-fermion system, that – inherently – is intricate to treat because of the "sign problem" that goes with it.

The richness and profundity of Landau–Pekar's polaron concept is further illustrated by its extensions to lattice polarons. Even the simplest two-site polaron model by Holstein (Sect. 3.1) proved to be very useful for a qualitative understanding of nontrivial features of the polaron problem, and for obtaining some novel analytical and semianalytical results (see, e.g., [138, 151]).

The "$1/\lambda$" expansion technique based on the Lang–Firsov transformation (Sect. 3.2.1) and unbiased numerical analysis of the finite and infinite Holstein and Fröhlich models combining Lanczos diagonalizations of clusters, density matrix renormalization group, cluster perturbation theory techniques, DMFT and different QMC algorithms, allowed for a description of properties of a single lattice polaron and a lattice bipolaron. The Lang–Firsov canonical transformation [19] was proven particularly instrumental in calculation of different kinetic and optical coefficients, which can be represented as expansions in powers of the unrenormalized hopping integral t (Sect. 3.8). Sometimes, it is possible to sum the expansion and get results, which are valid for arbitrary values of parameters providing the understanding of the crossover region from the Boltzmann kinetics to thermally activated hopping [234].

Recent ED [35], CTQMC [184], and DQMC [58] techniques allow for determination of the ground state and excited states of lattice polarons with arbitrary precision in the thermodynamic limit for any dimension and any type of lattices (Sect. 3.4.2). The spectral properties (e.g., photoemission), optical response and thermal transport, as well as the dynamics of polaron formation in the Holstein model have been numerically analyzed for all EPI strengths and phonon frequencies, including the intermediate-coupling regime (see Sect. 3.8 and [35]). CTQMC methods have proven to be powerful and versatile tools providing unbiased results for the polaron properties in any lattices for any-range EPI, including Jahn–Teller polarons (Sect. 3.4). Combining the Lang–Firsov transformation and quantum Monte Carlo simulations allows for an exact sampling without autocorrelations, which proves to be an enormous advantage for small phonon frequencies or low temperatures [177].

Importantly, variational and numerical techniques confirmed that the Fröhlich and Holstein–Lang–Firsov theories are asymptotically exact in the weak, $\lambda \ll 1$, and strong-coupling, $\lambda \gg 1$, regimes, respectively, and the polaron formation represents a continuous crossover of the ground state (Sect. 3.4). The crossover is related to the exponential increase of the effective mass, and the band narrowing with a strongly suppressed electronic quasiparticle residue and the Drude weight, accompanied by an increase of the incoherent spectral weight (Sect. 3.8.3). These features strongly depend on the phonon dispersion [141], EPI radius [59, 60, 269, 270], lattice geometry [37], and are more pronounced in higher dimensions [37, 269]. Remarkably, the unscreened Fröhlich EPI provides relatively light lattice polarons (Sect. 3.3), which are several orders of magnitude lighter than the Holstein small polarons [59, 60] at strong coupling. This classification of weak- and strong-coupling regimes still leaves room for Fröhlich polarons of not only weak but also intermediate and – in the theoretical analysis – of strong coupling classified with respect to the Fröhlich electron–phonon coupling constant α (Sect. 2.3).

While the single polaron has been actively researched for a long time and is now well understood, the multipolaron physics has gained particular attention in the last two decades. It has been found – unexpectedly for many researchers – that the Migdal–Eliashberg theory breaks down already at $\lambda \sim 1$ for any adiabatic ratio ω_0/E_F. The effective parameter $\lambda\omega_0/E_F$ becomes large at $\lambda \gtrsim 1$ since the bandwidth is narrowed and the Fermi energy, E_F, is renormalized down exponentially [125,130]. Extending the BCS theory toward the strong interaction between electrons and ion vibrations, a charged Bose gas of tightly bound small bipolarons was predicted [271], with a further prediction that the highest superconducting transition temperature is attained in the crossover region of EPI strength between the BCS-like polaronic (Sect. 5.3) and bipolaronic (Sect. 5.5) superconductivity [130]. Subsequent studies of the Holstein–Hubbard model found sufficiently mobile two-site bipolarons [183, 277, 288, 289] even in this model, generally unfavorable for tunneling. Taking into account that many advanced materials with low density of free carriers and poor mobility are characterized by poor screening of high-frequency optical phonons, the Coulomb–Fröhlich multipolaron lattice model was introduced and related to high-temperature cuprate superconductors [59,311]. The large Hubbard U and intersite Coulomb repulsions and the unscreened Fröhlich EPI provide "superlight" but small intersite bipolarons (Sect. 4.4). More recent CTQMC simulations of intersite small bipolarons in the Fröhlich–Coulomb model [312] have found such quasiparticles in a wide parameter range with achievable phonon frequencies and couplings. They could have a superconducting transition in excess of room temperature.

Although a rather complete understanding of the Fröhlich and Holstein polarons and bipolarons in the weak, crossover and strong-coupling regimes is now achieved in the dilute limit, the multipolaron problem remains a challenging one when EPI competes with sometimes strong electronic correlations (Sect. 5.6). The corresponding microscopic models contain (extended) Hubbard, Heisenberg, or double-exchange terms, and maybe also a coupling to orbital degrees of freedom along with strong EPI, so that even numerical solutions with the same precision as in the dilute (bi)polaron case are often problematic. A number of ED, QMC, DMFT, and combined numerical results give strong evidence that the tendency toward lattice polaron formation is enhanced in strongly correlated electron systems due to a narrowing of the electron band caused by strong correlations (see Sect. 5.6 and reviews [35, 176, 177, 218] for more details). Not only antiferromagnetic correlations enhance EPI, resulting in polaron formation for moderate coupling strength, but also EPI strongly enhances spin correlations [152, 517]. Some of these studies show that increasing carrier density could be accompanied by a dissociation of polarons, leading to normal metallic behavior in the intermediate-coupling adiabatic regime [176] that is reminiscent of the "overcrowding" effect hypothesized by Mott [518]. On the other hand for parameters favoring small polarons, no such density-driven crossover occurs in agreement with simple analytical results [125] (see Sect. 5.2). A more thorough investigation of these models will definitely be a great challenge in the near future.

However, exactly solvable models might give a rather limited, sometimes misleading, description of polarons in real systems. Qualitative inconsistencies can arise when coupling is assumed to be just to one phonon mode, often taken as dispersionless, and ad hoc approximations for EPI matrix elements are applied. Moreover, electronic nanoscale disorder and long-range strain fields can interweave with the microscopic mechanisms of polaronic transport [392]. Hence, ab initio calculations of the phonon spectrum, EPI, and polaron properties beyond the adiabatic Born–Oppenheimer approximation are required in many cases for which the theory and experiment can be compared in detail [519].

References

1. E.G. Maximov, D.Yu. Savrasov, S.Yu. Savrasov, Uspechi Fiz. Nauk. **167**, 353 (1997)
2. M. Born, R. Oppenheimer, Rev. Mod. Phys. **84**, 457 (1927)
3. P. Hohenberg, W Kohn, Proc. R. Soc. Lond. A **136**, 864 (1937)
4. W. Kohn, L.J. Sham, Phys. Rev. A **140**, 1133 (1965)
5. V.G. Baryakhtar, E.V. Zarochentsev, E.P. Troitskaya, *Theory of Adiabatic Potential and Atomic Properties of Simple Metals* (Gordon and Breach, Amsterdam, 1999)
6. D. Belitz, T.R. Kirkpatrick, Rev. Mod. Phys. **66**, 261 (1994)
7. G.D. Mahan, *Many-Particle Physics* (Kluwer/Plenum, New York, 1990)
8. G.D. Mahan, in *Polarons in Ionic Crystals and Polar Semiconductors*, ed. J.T. Devreese (North-Holland, Amsterdam, 1972), p. 553
9. A.S. Alexandrov, *Theory of Superconductivity: From Weak to Strong Coupling* (IoP, Bristol, 2003)
10. L.D. Landau, Physikalische Zeitschrift der Sowjetunion **3**, 664 (1933)
11. J.T. Devreese, in *Encyclopedia of Applied Physics*, vol. 14, ed. by G.L. Trigg (VCH, Weinheim, 1996), p. 383
    J.R. Schrieffer, M.L. Chiofalo (IOS Press, Amsterdam, 1998), p. 287
12. R.P. Feynman, Phys. Rev. **97**, 660 (1955)
13. H. Fröhlich, Adv. Phys. **3**, 325 (1954)
14. S.I. Pekar, Zh. Eksp. Teor. Fiz. **16**, 335 (1946)
15. E.I. Rashba, Optika i spektroskopia **2**, 75 (1957)
16. D.M. Eagles, Phys. Rev. **130**, 1381 (1963)
17. T. Holstein, Ann. Phys. **8**, 325 (1959)
18. T. Holstein, Ann. Phys. **8**, 343 (1959)
19. I.G. Lang, Y.A. Firsov, Zh. Eksp. Teor. Fiz. **43**, 1843 (1962) [Sov. Phys. JETP **16**, 1301 (1962)].
20. G.L. Sewell, Philos. Mag. **3**, 1361 (1958)
21. S.V. Tyablikov, Zh. Eksp. Teor. Fiz **23**, 381 (1952)
22. J. Yamashita, T. Kurosawa, J. Phys. Chem. Solids **5**, 34 (1958)
23. J. Devreese, Bull. Soc. Belge Phys., Ser. III **4**, 259 (1963)
24. P. Nagels, M. Denayer, J. Devreese, Solid State Commun. **1**, 35 (1963)
25. A.S. Alexandrov, N.F. Mott, Rep. Prog. Phys. **57**, 1197 (1994)

26. A.S. Alexandrov, N.F. Mott, *Polarons and Bipolarons* (World Scientific, Singapore, 1995)
27. J. Appel, in *Solid State Physics*, vol. 21, ed. by F. Seitz, D. Turnbull, H. Ehrenreich (Academic, London, 1968)
28. H. Boettger, V. Bryksin, *Hopping Conduction in Solids* (Academic, Berlin, 1985).
29. Y.A. Firsov, *Polarons* (Nauka, Moscow, 1975), p. 553
30. N. Itoh, A.M. Stoneham, *Materials Modification by Electronic Excitation* (Cambridge University Press, Cambridge, 2001)
31. T.K. Mitra, A. Chatterjee, S. Mukhopadhyay, Phys. Rep. **153**, 91 (1987)
32. E.I. Rashba, in *Encyclopedia of Condensed Matter Physics*, ed. by G.F. Bassani, G.L. Liedi, P. Wyder (Elsevier, Amsterdam, 2005), p. 347
33. E.K.H. Salje, A.S. Alexandrov, W.Y. Liang (eds.), *Polarons and Bipolarons in High-Tc Superconductors and Related Materials* (Cambridge University Press, Cambridge, 1995)
34. J.T. Devreese, A.S. Alexandrov, Rep. Prog. Phys. **72**, 066501 (2009)
35. H. Fehske, S.A. Trugman, in *Polarons in Advanced Materials*, ed. by A.S. Alexandrov (Canopus/Springer, Bristol, 2007), p. 393
36. B. Gerlach, H. Löwen, Rev. Mod. Phys. **63**, 63 (1991)
37. J.P. Hague, P.E. Kornilovitch, A.S. Alexandrov, J.H. Samson, Phys. Rev. B **73**, 054303 (2006)
38. H. Löwen, Phys. Rev. B **37**, 8661 (1988)
39. F.M. Peeters, J.T. Devreese, Physica Status Solidi (b) **112**, 219 (1982)
40. S.I. Pekar, *Research in Electron Theory of Crystals* (US AEC Transl. AEC-tr-555, Russian edition 1951, German edition 1954, 1951)
41. L.D. Landau, S.I. Pekar, Zh. Eksp. Teor. Fiz. **18**, 419 (1948)
42. H. Fröhlich, H. Pelzer, S. Zienau, Philos. Mag. **41**, 221 (1950)
43. T.D. Lee, D. Pines, F. Low, Phys. Rev. **90**, 297 (1953)
44. M. Gurari, Philos. Mag. **44**, 329 (1953)
45. T.D. Lee, D. Pines, Phys. Rev. **88**, 960 (1952)
46. G.R. Allcock, in *Polarons and Excitons*, ed. by C.G. Kuper, G.D. Whitfield (Oliver and Boyd, Edinburgh, 1963), p. 45
47. N.N. Bogoliubov, Ukr. Math. Zh. **2**, 3 (1950)
48. N.N. Bogoliubov, Jr., in *Superconductivity and Strongly Correlated Electron Systems*, ed. by C. Noce, A. Romano, G. Scarpetta (World Scientific, Singapore, 1994), p. 107
49. T.D. Schultz, in *Polarons and Excitons*, ed. by C.G. Kuper, G.D. Whitfield (Oliver and Boyd, Edinburgh, 1962), p. 71
50. R.P. Feynman, *Statistical Mechanics* (Benjamin, New York, 1972)
51. M.A. Krivoglaz, S.I. Pekar, Bull. Acad. Sci. U.S.S.R. **21**, 1, 13, 29 (1957)
52. Y. Osaka, Prog. Theor. Phys. **22**, 437 (1959)
53. W. Becker, B. Gerlach, H. Schliffke, Phys. Rev. B **28**, 5735 (1983)
54. F.M. Peeters, J.T. Devreese, Phys. Rev. B **31**, 6826 (1985)
55. A.S. Mishchenko, N.V. Prokof'ev, A. Sakamoto, B.V. Svistunov, Phys. Rev. B **62**, 6317 (1990)
56. N.V. Prokof'ev, B.V. Svistunov, Phys. Rev. Lett. **81**, 2514 (1998)
57. N. Metropolis, A.W. Rosenbluth, M.N. Rosenbluth, A.M. Teller, E. Teller, J. Chem. Phys. **21**, 1087 (1953)
58. A.S. Mishchenko, N. Nagaosa,in *Polarons in Advanced Materials*, ed. by A.S.Alexandrov(Canopus/Springer, Bristol,2007), p. 503

59. A.S. Alexandrov, Phys. Rev. B **53**, 2863 (1996)
60. A.S. Alexandrov, P.E. Kornilovitch, Phys. Rev. Lett. **82**, 807 (1999)
61. P.E. Spencer, J.H. Samson, P.E. Kornilovitch, A.S. Alexandrov, Phys. Rev. B **71**, 184310 (1995)
62. J.T. Devreese, (1964) *Contribution to the Polaron Theory, Ph.D. Thesis* (KU Leuven)
63. J.T. Devreese, R. Evrard, Phys. Lett. **11**, 278 (1964)
64. J.T. Devreese, R. Evrard, in *Proceedings of the British Ceramic Society* **10**, 151 (1968) [reprinted in *Path Integrals and Their Applications in Quantum, Statistical, and Solid State Physics*, NATO ASI Series B, Physics, vol. 34, ed. by G.J. Papadopoulos, J.T. Devreese (Plenum, New York, 1968), p. 344 (1977)]
65. J.T. Titantah, C. Pierleoni, S. Ciuchi, Phys. Rev. Lett. **87**, 206406 (2001)
66. V. Cataudella, G. De Filippis, C.A. Perroni, in *Polarons in Advanced Materials*, ed. by A.S. Alexandrov (Canopus/Springer, Bristol, 2007), p. 149
67. G. De Filippis, V. Cataudella, V. Marigliano Ramaglia, C.A. Perroni, D. Bercioux, Eur. Phys. J. B **36**, 65 (2003)
68. C. Grosjean, in *Lecture Notes on Solid State Physics*, ed. by D. Pines (Princeton University Press, Princeton, 1957), Chap. 8
69. G. Höhler, A.M. Mullensiefen, Z. Physik **157**, 159 (1959)
70. J. Röseler, Physica Status Solidi **25**, 311 (1968)
71. R.P. Feynman, R.W. Hellwarth, C.K. Iddings, P.M. Platzman, Phys. Rev. **127**, 1004 (1962)
72. K.K. Thornber, R.P. Feynman, Phys. Rev. B **1**, 4099 (1970)
73. J.T. Devreese, J. De Sitter, M. Goovaerts, Phys. Rev. B **5**, 2367 (1972)
74. F.M. Peeters, J.T. Devreese, Phys. Rev. B **34**, 7246 (1986)
75. Fröhlich, H., Proc. R. Soc. Lond. A **160**, 230 (1937)
76. D.J. Howarth, E.H. Sondheimer, Proc. R. Soc. Lond. A **219**, 53 (1953)
77. Y. Osaka, Prog. Theor. Phys. **25**, 517 (1961)
78. F.E. Low, D. Pines, Phys. Rev. **98**, 414 (1955)
79. D.C. Langreth, L.P. Kadanoff, Phys. Rev. **133**, A1070 (1964)
80. P.M. Platzman, in *Polarons and Excitons*, ed. by C.G. Kuper, G.D. Whitfield, (Oliver and Boyd, Edinburgh, 1963), p. 123
81. R.P. Feynman, Phys. Rev. **84**, 108 (1951)
82. F.M. Peeters, J.T. Devreese, Phys. Rev. B **28**, 6051 (1983)
83. F.M. Peeters, J.T. Devreese, in *Solid State Physics*, vol. 38, ed. by F. Seitz, D. Turnbull, (Academic, New York, 1984), p. 81
84. L.P. Kadanoff, Phys. Rev. **130**, 1364 (1963)
85. F.M. Peeters, J.T. Devreese, Physica Status Solidi (b) **115**, 539 (1983)
86. R.W. Hellwarth, I Biaggio, Phys. Rev. B **60**, 299 (1999)
87. F.C. Brown, in *Polarons and Excitons*, ed. by C.G. Kuper, G.D. Whitfield (Oliver and Boyd, Edinburgh, 1963), p. 323
88. F.C. Brown, in *Point Defects in Solids*, vol. 1, ed. by J.H. Crawford, L.M. Slifkin (Plenum, New York, 1972), p. 537
89. I. Biaggio, R.W. Hellwarth, J.P. Partanen, Phys. Rev. Lett. **78**, 891 (1997)
90. E. Hendry, F. Wang, J. Shan, T.F. Heinz, M. Bonn, Phys. Rev.B **69**, 081101(R) (2004)
91. V.L. Gurevich, I.G. Lang, Yu.A. Firsov, Fiz. Tverd. Tela (St. Petersburg) **4**, 1252 [Sov. Phys. Solid State **4**, 918 (1962)]
92. J. Tempere, J.T. Devreese, Phys. Rev. B **64**, 104504 (2001)

93. J.T. Devreese, in *Polarons in Ionic Crystals and Polar Semiconductors* (North-Holland, Amsterdam, 1972), p. 83
94. J.T. Devreese, in *Proceedings of the "Enrico Fermi" Summer School, Course CLXI – "Polarons in Bulk Materials and Systems with Reduced Dimensionality", Varenna, June 21–July 1, 2005*, ed. by G. Iadonisi, J. Ranninger, G. De Filippis (IOS Press, Amsterdam, 2006), p. 27
95. J. Devreese, W. Huybrechts, L. Lemmens, Physica Status Solidi (b) **48**, 77 (1971)
96. J.T. Devreese, J. Tempere, Solid State Commun. **106**, 309 (1998)
97. H. Finkenrath, N. Uhle, W. Waidelich, Solid State Commun. **7**, 11 (1969)
98. E. Kartheuser, E. Evrard, J. Devreese, Phys. Rev. Lett. **22**, 94 (1969)
99. M.J. Goovaerts, J. De Sitter, J.T. Devreese, Phys. Rev. **7**, 2639
100. A.S. Mishchenko, N. Nagaosa, N.V. Prokof'ev, A. Sakamoto, B.V. Svistunov, Phys. Rev. Lett. **91**, 236401 (2003)
101. G. De Filippis, V. Cataudella, A.S. Mishchenko, C.A. Perroni, J.T. Devreese, Phys. Rev. Lett. **96**, 136405 (2006)
102. J.T. Devreese, J. De Sitter, M. Goovaerts, Solid State Commun. **9**, 1383 (1971)
103. W. Huybrechts, J. Devreese, Solid State Commun. **17**, 401 (1975)
104. D.M. Eagles, R.P.S.M. Lobo, F. Gervais, Phys. Rev. B **52**, 6440 (1995)
105. J.T. Devreese, in *Lectures on the Physics of Highly Correlated Electron Systems VII*, ed. by A. Avella, F. Mancini (AIP, Melville, 2003), p. 3
106. J.W. Hodby, G.P. Russell, F.M. Peeters, J.T. Devreese, D.M. Larsen, Phys. Rev. Lett. **58**, 1471 (1987)
107. N. Miura, H. Nojiri, P. Pfeffer, W. Zawadzki, Phys. Rev. B **55**, 13598 (1997)
108. J.M. Shi, F.M. Peeters, J.T. Devreese, Phys. Rev. B **48**, 5202 (1993)
109. D. Larsen, J. Phys. C **7**, 2877 (1974)
110. Cheng, J.-P., B.D. McCombe, J.M. Shi, F.M. Peeters, J.T. Devreese, Phys. Rev. B **48**, 7910 (1993)
111. W. Huybrechts, J. Devreese, Phys. Rev. B **8**, 5754 (1973)
112. J.T. Devreese, S.N. Klimin, (2006), unpublished data
113. W. Götze, P. Wolfle, Phys. Rev. B **6**, 1226 (1972)
114. H. Mori, Prog. Theor. Phys. **33**, 423 (1965)
115. J.T. Devreese, in *Polarons in Advanced Materials*, ed. by A.S. Alexandrov (Canopus/Springer, Bristol, 2007), p. 3
116. S.J. Miyake, J. Phys. Soc. Jap. **38**, 181 (1975)
117. G.R. Allcock, Adv. Phys. **5**, 412 (1956)
118. J.T. Devreese, L. Lemmens, J. Van Royen, Phys. Rev. B **15**, 1212 (1977)
119. L.F. Lemmens, J. De Sitter, J.T. Devreese, Phys. Rev. B **8**, 2717 (1973)
120. F.M. Peeters, W. Xiaoguang, J.T., Devreese, Phys. Rev. B **33**, 3926 (1986)
121. F.M. Peeters, J.T. Devreese, Phys. Rev. B **36**, 4442 (1987)
122. W. Xiaoguang, F.M. Peeters, J.T. Devreese, Phys. Rev. B **31**, 3420 (1985)
123. A.S. Alexandrov, A.M. Bratkovsky, Phys. Rev. Lett. **84**, 2043 (2000)
124. D.M. Eagles, Phys. Rev. **145**, 645 (1963)
125. A.S. Alexandrov, Europhys. Lett. **56**, 92 (2001)
126. A.S. Alexandrov, V.V. Kabanov, D.K. Ray, Phys. Rev. B **49**, 9915 (1994)
127. A.O. Caldeira, A.J. Leggett, Phys. Rev. Lett. **46**, 211 (1981)
128. A.H. Castro Neto, A.O. Caldeira, Phys. Rev. Lett. **67**, 1960 (1991)
129. I.G. Lang, Y.A. Firsov, Zh. Eksp. Teor. Fiz. **45**, 378 (1963) [Sov. Phys. JETP **18**, 262 (1963)].

130. A.S. Alexandrov, Zh. Fiz. Khim. **57**, 273 (1983) [Russ. J. Phys. Chem. **57**, 167]
131. A.S. Alexandrov, Phys. Rev. B **46**, 2838 (1992)
132. A.S. Alexandrov, H. Capellmann, Phys. Rev. B **43**, 2042 (1991)
133. A.S. Alexandrov, H. Capellmann, U. Göbel, Phys. Rev. B **46**, 4374 (1992)
134. U. Gobel, A.S. Alexandrov, H. Capellmann, Zeitschrift Physik B **96**, 47 (1994)
135. H. Rietschel, L. Pintschovius, W. Reichardt, Physica C (Amsterdam) **162**, 1705 (1989)
136. A.S. Alexandrov, Phys. Rev. B **61**, 12315 (2000)
137. A.A. Gogolin, Physica Status Solidi (b) **109**, 95 (1982)
138. E.K. Kudinov, Yu.A. Firsov, Fiz. Tverd. Tela (St. Petersburg) **39**, 2159 (1997) [Phys. Solid State **39**, 1930 (1997)].
139. E.V.L. De Mello, J. Ranninger, Phys. Rev. B **58**, 9098 (1998)
140. A.S. Alexandrov, B.Y. Yavidov, Phys. Rev. B **69**, 073101 (2004)
141. M. Zoli, Phys. Rev. B **57**, 10555 (1998)
142. M. Zoli, Phys. Rev. B **61**, 14523 (1998)
143. M. Zoli, A.N. Das, J. Phys. Condens. Matter **16**, 3597 (2004)
144. J. Chatterjee, A.N. Das, Phys. Rev. B **61**, 4592 (2000)
145. J. Chatterjee, A.N. Das, P. Choudhury, Solid State Commun. **126**, 113 (2003)
146. D.M. Eagles, Phys. Rev. **181**, 1278 (1969)
147. D. Emin, T. Holstein, Phys. Rev. Lett. **36**, 323 (1976)
148. G. Kalosakas, S. Aubry, G.P. Tsironis, Phys. Rev. B **58**, 3094 (1998)
149. H. Zheng, M. Avignon, Phys. Rev. B **68**, 024301 (2003)
150. Z. Hang, Phys. Rev. B **37**, 7419 (1988)
151. R.S. Han, Z.J. Lin, K.L. Wang, Phys. Rev. B **65**, 174303 (2002)
152. M. Capone, S. Ciuchi, Phys. Rev. B **65**, 104409 (1997)
153. S. Swain, J. Phys. A **6**, 192 (1973)
154. M. Berciu, Phys. Rev. B **75**, 081101(R) (2007)
155. O.S. Barisic, Phys. Rev. B **69**, 064302 (2004)
156. M. Capone, W. Stephan, M. Grilli, Phys. Rev. B **56**, 4484 (1997)
157. H. Fehske, H. Röder, G. Wellein, A. Mistriotis, Phys. Rev. B **51**, 16582 (1995)
158. A. Kongeter, M. Wagner, J. Chem. Phys. **92**, 4003 (1990)
159. F. Marsiglio, Phys. Lett. A **180**, 280 (1993)
160. F. Marsiglio, Physica C **244**, 21 (1995)
161. J. Ranninger, U. Thibblin, Phys. Rev. B **45**, 7730 (1992)
162. W. Stephan, Phys. Rev. B **54**, 8981 (1996)
163. G. Wellein, H. Röder, H. Fehske, Phys. Rev. B **53**, 9666 (1996)
164. A.H. Romero, D.W. Brown, K. Linderberg, J. Chem. Phys. **109**, 6540 (1998)
165. L.C. Ku, S.A. Trugman, J. Bonča, Phys. Rev. B **65**, 174306 (2002)
166. S.A. Trugman, J. Bonča, L.C. Ku, Int. J. Mod. Phys. B **15**, 2707 (2001)
167. E. Berger, P. Valášek, W. von der Linden, Phys. Rev. B **52**, 4806 (1995)
168. H. De Raedt, A. Lagendijk, Phys. Rev. Lett. **49**, 1522 (1982)
169. H. De Raedt, A. Lagendijk, Phys. Rev. B **27**, 6097 (1983)
170. H. De Raedt, A. Lagendijk, Phys. Rev. B **30**, 1671 (1984)
171. H. De Raedt, A. Lagendijk, Phys. Rep. **127**, 234 (1985)
172. E. Fradkin, J.E. Hirsch, Phys. Rev. B **27**, 1680 (1983)
173. J.E. Hirsch, E. Fradkin, Phys. Rev. Lett. **49**, 402 (1982)
174. J.E. Hirsch, E. Fradkin, Phys. Rev. B **27**, 4302 (1983)
175. M. Hohenadler, H.G. Evertz, W. von der Linden, Phys. Rev. B **69**, 024301 (2004)

176. M. Hohenadler, D. Neuber, W. von der Linden, G. Wellein, J. Loos, H. Fehske, Phys. Rev. B **71**, 245111 (2005)
177. M. Hohenadler, W. von der Linden, in *Polarons in Advanced Materials*, ed. by A.S. Alexandrov (Canopus/Springer, Bristol, 2007), p. 463
178. E. Jeckelmann, S.R. White, Phys. Rev. B **57**, 6376 (1998)
179. E. Jeckelmann, C. Zhang, S.R. White, Phys. Rev. B **60**, 7950 (1999)
180. C. Zhang, E. Jeckelmann, S.R. White, Phys. Rev. B **60**, 14092 (1999)
181. P.E. Kornilovitch, Phys. Rev. Lett. **81**, 5382 (1998)
182. P.E. Kornilovitch, Phys. Rev. B **60**, 3237 (1999)
183. A. Macridin, G.A. Sawatzky, M. Jarrell, Phys. Rev. B **69**, 245111 (2004)
184. P.E. Kornilovitch, in *Polarons in Advanced Materials*, ed. by A.S. Alexandrov (Canopus/Springer, Bristol, 2007), p. 191
185. E.V.L. De Mello, J. Ranninger, Phys. Rev. B **55**, 14872 (1997)
186. Y.A. Firsov, V.V. Kabanov, E.K. Kudinov, A.S. Alexandrov, Phys. Rev. B **59**, 12132 (1999)
187. E.V.L. De Mello, J. Ranninger, Phys. Rev. B **58**, 14625 (1998)
188. E.V.L. De Mello, J. Ranninger, Phys. Rev. B **59**, 12135 (1999)
189. H. Fehske, J. Loos, G. Wellein, Z. Phys. B **104**, 619 (1997)
190. J. Bonča, S.A. Trugman, I. Batistić, Phys. Rev. B **60**, 1633 (1999)
191. P.E. Kornilovitch, E.R. Pike, Phys. Rev. B **55**, R 8634 (1997)
192. N.A. Perroni, V. Cataudella, G. De Filippis, J. Phys. Condens. Matter **16**, 1593 (2004)
193. A.H. Romero, D.W. Brown, K. Lindenberg Phys. Rev. B **59**, 13728 (1999)
194. G. Wellein, H. Fehske Phys. Rev. B **56**, 4513 (1997)
195. Y.B. Levinson, E.I. Rashba, Rep. Prog. Phys. **36**, 1499 (1973)
196. A.S. Alexandrov, Phys. Rev. B **46**, 14932 (1992)
197. R. Khasanov, D.G. Eshchenko, H. Luetkens, E. Morenzoni, T. Prokscha, A. Suter, N. Garifianov, M. Mali, J. Roos, K. Conder, H. Keller, Phys. Rev. Lett. **92**, 057602 (2004)
198. G.M. Zhao, D.E. Morris, Phys. Rev. B **51**, 16487 (1995)
199. G.-M. Zhao, M.B. Hunt, H. Keller, K.A. Müller, Nature (London) **385**, 236 (1997)
200. A.S. Alexandrov, G.M. Zhao, H. Keller, B. Lorenz, Y.S. Wang, C.W. Chu, Phys. Rev. B **64**, 140404(R) (2001)
201. G.-H. Gweon, T. Sasagawa, S.Y. Zhou, J. Craf, H. Takagi, D.H. Lee, A. Lanzara, Nature (London) **430**, 187 (2004)
202. A. Deppeler, A.J. Millis, Phys. Rev. B **65**, 224301 (2002)
203. P. Paci, M. Capone, E. Cappelluti, S. Ciuchi, C. Grimaldi, L. Pietronero, Phys. Rev. Lett. **94**, 036406 (2005)
204. P.E. Kornilovitch, A.S. Alexandrov, Phys. Rev. B **70**, 224511 (2004)
205. A. Bussmann-Holder, H. Keller, A.R. Bishop, A. Simon, R. Micnas, K.A. Muller, Europhys. Lett. **72**, 423 (2005)
206. S. Fratini, S. Ciuchi, Phys. Rev. B. **72**, 235107 (2005)
207. A.S. Mishchenko, N. Nagaosa, Phys. Rev. B **73**, 092502 (2006)
208. W.P. Su, J.R. Schrieffer, A.J. Heeger, Phys. Rev. Lett. **42**, 1698 (1979)
209. M. Zoli, in *Polarons in Advanced Materials*, ed. by A.S. Alexandrov (Canopus/Springer, Bristol, 2007), p. 231
210. H.A. Jahn, E. Teller, Phys. Rev. B **161**, 220 (1937)
211. J.G. Bednorz, K.A. Müller, Rev. Mod. Phys. **60**, 585 (1988)

212. A.S. Alexandrov, V.V. Kabanov, Phys. Rev. B **54**, 3655 (1996)
213. O. Gunnarsson, Rev. Mod. Phys. **69**, 575 (1997)
214. T. Mertelj, V.V. Kabanov, D. Mihailović, Phys. Rev. Lett. **94**, 147003 (2005)
215. D. Mihailović, V.V. Kabanov, Phys. Rev. B **63**, 054505 (2001)
216. K.A. M üller, Physica C **341**, 11 (2000)
217. A. Bussmann-Holder, A.R. Bishop, Philos. Mag. B **79**, 119 (1999)
218. D.M. Edwards, Adv. Phys. **51**, 1259 (2002)
219. A.J. Millis, P.B. Littlewood, B.I. Shraiman, Phys. Rev. Lett. **74**, 5144 (1995)
220. H. Roder, J. Zang, A.R. Bishop, Phys. Rev. Lett. **76**, 1356 (1996)
221. S. Yunoki, A. Moreo, E. Dagotto, Phys. Rev. Lett. **81**, 5612 (1998)
222. J. Kanamori, J. Appl. Phys. **31**, S14 (1960)
223. P.B. Allen, V. Perebeinos, Phys. Rev. B **60**, 10747 (1999)
224. Y. Takada, Phys. Rev. B **61**, 8631 (2000)
225. P.E. Kornilovitch, Phys. Rev. Lett. **84**, 1551 (2000)
226. S. El Shawish, J. Bonča, L.C. Ku, S.A. Trugman, Phys. Rev. B **67**, 014301 (2003)
227. E.N. Economou, O. Yanovitskii, Th. Fraggis, Phys. Rev. B **47**, 740 (1993)
228. F.X. Bronold, A. Saxena, A.R. Bishop, Phys. Rev. B **63**, 235109 (2001)
229. J.P. Hague, P.E. Kornilovitch, A.S. Alexandrov, Phys. Rev. B **78**, 092302 (2008)
230. A.F. Ioffe, Can. Phys. **26**, 582 (1957)
231. M.I. Klinger, *Problems of Polaron Transport Theory in Semiconductors* (Pergamon Press, Oxford, 1979)
232. I.G. Lang, Y.A. Firsov, Zh. Eksp. Teor. Fiz. **54**, 826 (1968) [Sov. Phys. JETP **27**, 443 (1968)].
233. N.F. Mott, E.A. Davis, *Electronic Processes in Non-Crystalline Materials*, 2nd edn. (Oxford University Press, Oxford, 1979)
234. Y.A. Firsov, in *Polarons in Advanced Materials*, ed. by A.S. Alexandrov (Canopus/Springer, Bristol, 2007), p. 63
235. E.K. Kudinov, Yu.A. Firsov, Zh. Eksp. Teor. Fiz. **49**, 867 (1965) [Sov. Phys. JETP **22**, 603 (1965)].
236. E.K. Kudinov, Yu.A. Firsov, Fiz. Tverd. Tela (St. Petersburg) **8**, 666 (1966) [Sov. Phys. Solid State **8**, 536 (1966)].
237. R.E. Peierls, Z. Phys. **80**, 763 (1933)
238. H. Böttger, V.V. Bryksin, T. Damker in *Polarons in Advanced Materials*, ed. by A.S. Alexandrov (Canopus/Springer, Bristol, 2007), p. 107
239. V.V. Bryksin, Yu.A. Firsov, Sov. Phys. Solid State **12**, 627 (1970)
240. D. Emin, Ann. Phys. **64**, 336 (1971)
241. L. Friedman, T. Holstein, Ann. Phys. **21**, 494 (1963)
242. O. Entin-Wohlman, A.G. Aronov, Y. Levinson, Y. Imry, Phys. Rev. Lett. **75**, 4094 (1995)
243. A.A. Burkov, L. Balents, Phys. Rev. Lett. **91**, 057202 (2003)
244. O. Entin-Wohlman, A. Aharony, Y.M. Galperin, V.I. Kozub, V. Vinokur, Phys. Rev. Lett. **95**, 086603 (2005)
245. M.I. Klinger, Phys. Lett. **7**, 102 (1963)
246. H.G. Reik, Solid State Commun. **1**, 67 (1963)
247. D. Emin, Phys. Rev. B **48**, 13691 (1993)
248. P. Calvani, *Optical Properties of Polarons* (Editrice Compositori, Bologna, 2001)

249. D. Emin, Adv. Phys. **22**, 57 (1973)
250. H.G. Reik, in *Polarons in Ionic Crystals and Polar Semiconductors*, ed. by J.T. Devreese (North-Holland, Amsterdam, 1972), p. 679
251. P. Benedetti, R. Zeyher, Phys. Rev. B **58**, 14320 (1998)
252. S. Ciuchi, F. de Pasquale, S. Fratini, D. Feinberg, Phys. Rev. B **56**, 4494 (1997)
253. S. Fratini, S. Ciuchi, Phys. Rev. Lett. **91**, 256403 (2003)
254. H. Sumi, J. Phys. Soc. Jpn. **36**, 770 (1974)
255. S. Fratini, S. Ciuchi, Phys. Rev. B. **74**, 075101 (2006)
256. A.S. Alexandrov, V.V. Kabanov, D.K. Ray Physica C (Amsterdam) **224**, 247 (1994)
257. G. Wellein, H. Fehske, Phys. Rev. B **58**, 6208 (1998)
258. A. Weiße, G. Wellein, A. Alvermann, H. Fehske, Rev. Mod. Phys. **78**, 275 (2006)
259. G. Schubert, G. Wellein, A. Weiße, A. Alvermann, H. Fehske, Phys. Rev. B **72**, 104304 (2005)
260. E.K. Kudinov, D.N. Mirlin, Yu.A. Firsov, Fiz. Tverd. Tela (St. Petersburg) **11**, 2789 (1969)
261. A.S. Alexandrov, J. Ranninger, Phys. Rev. B **45**, 13109 (1992)
262. A.S. Alexandrov, C. Sricheewin, Europhys. Lett. **51**, 188 (2000)
263. G.M. Eliashberg, Zh. Eksp. Teor. Fiz. **38**, 966 (1960) [Sov. Phys. JETP **11**, 696 (1960)]
264. A.B. Migdal, 1958, Zh. Eksp. Teor. Fiz. **34**, 1438 [Sov. Phys. JETP **7**, 996 (1958)]
265. P.E. Kornilovitch, Europhys. Lett. **59**, 735 (2002)
266. M. Berciu, Phys. Rev. Lett. **97**, 036402 (2006)
267. M. Berciu, Phys. Rev. B **74**, 245104 (2006)
268. J. Loos, M. Hohenadler, H. Fehske, J. Phys. Condens. Matter **18**, 2453 (2006)
269. H. Fehske, J. Loos, G. Wellein, Phys. Rev. B **61**, 8016 (2002)
270. C.A. Perroni, V. Cataudella G. De Filippis, V.M. Ramaglia, Phys. Rev. B **71**, 054301 (2005)
271. A. Alexandrov, J. Ranninger, Phys. Rev. B **23**, 1796 (1981)
272. A. Alexandrov, J. Ranninger, Phys. Rev. B **24**, 1164 (1981)
273. P.W. Anderson, Phys. Rev. Lett. **34**, 953 (1975)
274. S. Aubry, J. Phys. (France) IV Colloq. C2 **3**, 349 (1993)
275. G. Beni, P. Pincus, J. Kanamori, Phys. Rev. B **10**, 1896 (1974)
276. R.A. Street, N.F. Mott, Phys. Rev. Lett. **1293**, 1975 (1975)
277. S. Aubry, in *Polarons and Bipolarons in High Tc Superconductors and Related Materials*, ed. by E.K.H. Salje, A.S. Alexandrov, W.Y. Liang (Cambridge University Press, Cambridge, 1995), p. 271
278. A.S. Alexandrov, V.V. Kabanov, Fiz. Tverd. Tela (St. Petersburg) **28**, 1129 (1986) [Sov. Phys. Solid State **28**, 631]
279. V.V. Bryksin, A.V. Gol'tsev, Fiz. Tverd. Tela (St. Petersburg) **30**, 1476 (1988) [Sov. Phys. Solid State **30**, 851 (1988)]
280. V.V. Bryksin, V.S. Voloshin, Fiz. Tverd. Tela (St. Petersburg) **26**, 2357 (1984) [Sov. Phys. Solid State **26**, 1429 (1984)]
281. A.S. Alexandrov, A.M. Bratkovsky, J. Phys. Condens. Matter **48**, L531 (1999)
282. M. Hohenadler, M. Aichhorn, W. von der Linden, Phys. Rev. B **71**, 014302 (2005)
283. S. Aubry, in *Polarons in Advanced Materials*, ed. by A.S. Alexandrov (Canopus/Springer, Bristol, 2007), p. 311

284. L. Proville, S. Aubry, Physica D **113**, 307 (1998)
285. L. Proville, S. Aubry, Eur. Phys. J. B **11**, 41 (1999)
286. A. Alexandrov, Physica C (Amsterdam) **182**, 327 (1991)
287. S. Sil, J. Phys. Condens. Matter **11**, 8879 (1999)
288. J. Bonča, T. Katrasnic, S.A. Trugman, Phys. Rev. Lett. **84**, 3153 (2000)
289. A. La Magna, R. Pucci, Phys. Rev. B **55**, 14886 (1997)
290. S. Robaszkiewicz, R. Micnas, K.A. Chao, Phys. Rev. B **23**, 1447 (1981)
291. T. Hakioglu, M.Y. Zhuravlev, Phys. Rev. B **58**, 3777 (1998)
292. W. Koller, D. Meyer, Y. Ono, A.C. Hewson, Europhys. Lett. **66**, 559 (2004)
293. P.W. Anderson, *The Theory of Superconductivity in the Cuprates* (Princeton University Press, Princeton, NJ, 1997)
294. B.K. Chakraverty, J. Ranninger, D. Feinberg, Phys. Rev. Lett. **81**, 433 (1998)
295. J. Adamowski, Phys. Rev. B **39**, 3649 (1989)
296. F. Bassani, M. Geddo, G. Iadonisi, D. Ninno, Phys. Rev. B **43**, 5296 (1991)
297. H. Hiramoto, Y. Toyozawa, J. Phys. Soc. Jpn. **54**, 245 (1985)
298. E.A. Kochetov, S.P. Kuleshov, V.A. Mateev, M.A. Smondyrev, Teor. Mat. Fiz. **30**, 183 (1977)
299. V.K. Mukhomorov, Fiz. Tekh. Poluprovodn. (St. Peterburg) **16**, 1095 (1982) [Sov. Phys. Semicond. **16**, 700 (1982)]
300. S.G. Suprun, B.Y. Moizhes, Fiz. Tverd. Tela (St. Peterburg) **24**, 1571 (1982)
301. G. Verbist, F.M. Peeters, J.T. Devreese, Solid State Commun. **76**, 1005 (1990)
302. G. Verbist, F.M. Peeters, J.T. Devreese, Phys. Rev. B **43**, 2712 (1991)
303. V.L. Vinetskii Zh. Eksp. Teor. Fiz. **40**, 1459 (1961) [Sov. Phys. JETP **13**, 1023]
304. V.L. Vinetskii, M.S. Giterman, Zh. Eksp. Teor. Fiz. **33**, 730 (1957)
305. M.A. Smondyrev, V.M. Fomin, in *Polarons and Applications (Proceedings in Nonlinear Science)*, ed. by V.D. Lakhno (John Wiley & Sons, Chichester, 1994), p. 13
306. J.T. Devreese, in *Proceedings of the International School of Physics "Enrico Fermi", Course CXXXVI, Varenna, 1997, "Models and Phenomenology for Conventional and High-Temperature Superconductivity"*, ed. by G. Iadonisi,
307. L. Genzel, A. Wittlin, M. Bauer, M. Cardona, E. Schönherr, A. Simon, Phys. Rev. B **40**, 2170 (1989)
308. W. Kress, U. Schröder, J. Prade, A.D. Kulkarni, F.W. de Wette, Phys. Rev. B **38**, 2906 (1988)
309. M.A. Smondyrev, J.T. Devreese, F.M. Peeters, Phys. Rev. B **51**, 15008 (1995)
310. J.T. Devreese, G. Verbist, F. M. Peeters, in *Polarons and Bipolarons in High-Tc Superconductors and Related Materials*, ed. by E.K.H. Salje, A.S. Alexandrov, W.Y. Liang (Cambridge University Press, Cambridge, 1995), p. 385
311. A.S. Alexandrov, P.E. Kornilovitch J. Phys. Condens. Matter **14**, 5337 (2002)
312. J.P. Hague, P.E. Kornilovitch, J.H. Samson, A.S. Alexandrov, Phys. Rev. Lett. **98**, 037002 (2007)
313. J. Bonča, S.A. Trugman, Phys. Rev. B **64**, 094507 (2001)
314. G. Iadonisi, F. Bassani, Il Nuovo Cimento **2D**, 1541–1560 (1983)
315. G. Iadonisi, F. Bassani, Il Nuovo Cimento **3D**, 408–420 (1984)
316. G. Iadonisi, F. Bassani, Il Nuovo Cimento **9D**, 703–714 (1987)
317. R. Evrard, E. Kartheuser J. Devreese, Physica Status Solidi **41**, 431–438 (1970)
318. G. Iadonisi, G. Strinati F. Bassani, Physica Status Solidi (b) **153**, 611–622 (1989)
319. E. Burovski, H. Fehske, A.S. Mishchenko, Phys. Rev. Lett. **101**, 116403 (2008)

320. L.F. Lemmens, F. Brosens, D.J. T., Physica Status Solidi (b) **82**, 439 (1977)
321. S. Tomonaga, Prog. Theor. Phys. **2**, 6 (1947)
322. S. Fratini, P. Quémerais, Eur. Phys. J. B **14**, 99 (2000)
323. S.A. Jackson, P.M. Platzman, Phys. Rev. B **24**, 499 (1981)
324. V.B. Shikin, Y.P. Monarkha, Zh. Eksp. Teor. Fiz. **65**, 751 (1973) [Sov. Phys. JETP **38**, 373, (1973)].
325. U. Albrecht, P. Leiderer, Europhys. Lett. **3**, 705 (1987)
326. A.P. Volodin, M.S. Khaikin, V.S. Edelman, JETP Lett. **26**, 543 (1977)
327. M.M. Salomaa, G.A. Williams, Phys. Rev. Lett. **47**, 1730 (1981)
328. V.B. Shikin, JETP Lett. **27**, 39 (1978)
329. I.F. Silvera, Bull. Am. Phys. Soc. **46**, 1016 (2001)
330. J. Tempere, I.F. Silvera, J.T. Devreese, Phys. Rev. Lett. **87**, 275301 (2001)
331. J. Tempere, S.N. Klimin, I.F. Silvera, J.T. Devreese, Eur. Phys. J. B **32**, 329 (2003)
332. C.C. Grimes, G. Adams, Phys. Rev. Lett. **42**, 795 (1979)
333. V.M. Bedanov, F.M. Peeters, Phys. Rev. B **49**, 2667 (1994)
334. L.P. Gor'kov, D.M. Chernikova, Pis'ma Zh. Eksp. Teor. Fiz. **18**, 119 (1973) [JETP Lett. **18**, 68, (1973)]
335. J. Tempere, I.F. Silvera, J.T. Devreese, Phys. Rev. B **67**(3), 035402 (2003) (pages 8)
336. D.S. Fisher, B.I. Halperin, P.M. Platzman, Phys. Rev. Lett. **42**, 798 (1979)
337. G. Deville, A. Valdes, E.Y. Andrei, F.I.B. Williams, Phys. Rev. Lett. **53**, 588 (1984)
338. S. Lupi, P. Maselli, M. Capizzi, P. Calvani, P. Giura, P. Roy, Phys. Rev. Lett. **83**, 4852 (1999)
339. C. Hartinger, F. Mayr, J. Deisenhofer, A. Loidl, T. Kopp, Phys. Rev. B **69**, 100403(R) (2004)
340. J.L. van Mechelen, D. van der Marel, C. Grimaldi, A.B. Kuzmenko, N.P. Armitage, N. Reyren, H. Hagemann, I.I. Mazin, Phys. Rev. Lett. **100**, 226403 (2008)
341. V. Cataudella, G. De Filippis, G. Iadonisi, Eur. Phys. J. B **12**, 17 (1999)
342. J. Bardeen, L.N. Cooper, J.R. Schrieffer, Phys. Rev. **108**, 1175 (1957)
343. D.J. Scalapino, in *Superconductivity*, vol. 1, ed. by R.D. Parks (Marcel Deccer, NY, 1969), p. 449
344. A.S. Alexandrov, V.N. Grebenev, E.A. Mazur, Pis'ma Zh. Eksp. Teor. Fiz. **45**, 357 (1987) [JETP Lett. **45**, 455]
345. F. Dogan, F. Marsiglio, Phys. Rev. B **68**, 165102 (2003)
346. J.P. Hague, J. Phys. Condens. Matter **15**, 2535 (2003)
347. L. Pietronero, S. Strässler, C. Grimaldi, Phys. Rev. B **52**, 10516 (1995)
348. T. Frank, M. Wagner, Phys. Rev. B **60**, 3252 (1999)
349. Y. Takada, T. Higuchi, Phys. Rev. B **52**, 12720 (1995)
350. V.V. Kabanov, O.Y. Mashtakov, Phys. Rev. B **47**, 6060 (1993)
351. A.S. Alexandrov, V.V. Kabanov, JETP Lett. **72**, 569 (2000)
352. F.V. Kusmartsev, J. Physique (France) IV **9**, 321 (1999)
353. C.R.A. Catlow, M.S. Islam, J. Zhang, Phys. Condens. Matter **10**, L49 (1998)
354. A. Bianconi, N.L. Saini (eds.), *Stripes and Related Phenomena* (Kluwer Academic/Plenum, New York, 2000)
355. V.V. Kabanov, in *Polarons in Advanced Materials*, ed. by A.S. Alexandrov (Canopus/Springer, Bristol, 2007), p. 373

356. S.W. Cheong, H.Y. Hwang, C.H. Chen, B. Batlogg, L.W. Rupp, Jr., S.A. Carter, Phys. Rev. B **49**, 7088 (1994)
357. J.M. Tranquada, J.E. Lorenzo, D.J. Buttrey, V. Sachan, Phys. Rev. B **52**, 3581 (1995)
358. R.J. Bursill, R.H. McKenzie, J. Hamer, Phys. Rev. Lett. **80**, 5607 (1998)
359. M. Tezuka, R. Arita, H. Aoki, Phys. Rev. Lett. **95**, 226401 (2005)
360. M. Hohenadler, G. Wellein, A.R. Bishop, A. Alvermann, H. Fehske, Phys. Rev. B **73**, 245120 (2006)
361. H. Fehske, G. Wellein, G. Hager, A. Weiße, A.R. Bishop, Phys. Rev. B **69**, 165115 (2004)
362. R.T. Clay, R.P. Hardikar, Phys. Rev. Lett. **95**, 096401 (2005)
363. M. Capone, S. Ciuchi, Phys. Rev. Lett. **91**, 186405 (2003)
364. A.S. Alexandrov, D.A. Samarchenko, S.V. Traven, Zh. Eksp. Teor. Fiz. **93**, 1007 (1987) [Sov. Phys. JETP **66**, 567]
365. M.R. Schafroth, Phys. Rev. **100**, 463 (1955)
366. A.F. Andreev, I.M. Lifshitz, Sov. Phys. JETP **29**, 1107 (1969)
367. A.S. Alexandrov, J. Ranninger, S. Robaszkiewicz, Phys. Rev. B **33**, 4526 (1986)
368. G.G. Batrouni, R.T. Scalettar, Phys. Rev. Lett. **84**, 1599 (2000)
369. K. Kubo, S. Takada J. Phys. Soc. Jpn. **52**, 2108 (1983)
370. A.S. Moskvin, Phys. Rev. B **69**, 214505 (2004)
371. E. Dagotto, Rev. Mod. Phys. **66**, 763 (1994)
372. E. Dagotto, Science **309**, 257 (2005)
373. Yu.A. Izyumov, Yu.N. Skryabin, Phys. Usp. **44**, 109 (2001)
374. P.A. Lee, N. Nagaosa, X.G. Wen, Rev. Mod. Phys. **78**, 17 (2006)
375. E. Manousakis, Rev. Mod. Phys. **63**, 1 (1991)
376. N.M. Plakida, in *Lectures on the Physics of Highly Correlated Electron Systems*, vol. 580,ed. by F. Mancini (American Institute of Physics, Melville NY, 2001), 121
377. A.S. Alexandrov, A.M. Bratkovsky, Phys. Rev. Lett. **82**, 141 (1999)
378. P. Calvani, M. Capizzi, S. Lupi, P. Maselli, A. Paolone, R. Roy, S.W. Cheong, W. Sadowski, E. Walker, Solid State Commun. **91**, 113 (1994)
379. B.J. Campbell, R. Osborn, D.N. Argyriou, L. Vasiliu-Doloc, J.F. Mitchell, S.K. Sinha, U. Ruett, C.D. Ling, Z. Islam, J.W. Lynn, Phys. Rev. B **65**, 014427 (2001)
380. B.J. Campbell, S.K. Sinha, R. Osborn, S. Rosenkranz, J.F. Mitchell D.N. Argyriou, L. Vasiliu-Doloc, O.H. Seeck, J.W. Lynn, Phys. Rev. B **67**, 020409 (2003)
381. J.P. Falck, A. Levy, M.A. Kastner, R.J. Birgenau, Phys. Rev. B **48**, 4043 (1993)
382. A. Lanzara, P.V. Bogdanov, X.J. Zhou, S.A. Kellar, D.L. Feng, E.D. Lu, T. Yoshida, H. Eisaki, A. Fujimori, K. Kishio, J.I. Shimoyana, T. Noda, S. Uchida, Z. Hussain, Z.X. Shen, Nature (London) **412**, 510 (2001)
383. W. Meevasana, N.J.C. Ingle, D.H. Lu, J.R. Shi, F. Baumberger, K.M. Shen, W.S. Lee, T. Cuk, H. Eisaki, T.P. Devereaux, N. Nagaosa, J. Zaanen Z.X. Shen, Phys. Rev. Lett. **96**, 157003 (2005)
384. D. Mihailović, C.M. Foster, K. Voss, A.J. Heeger, Phys. Rev. B **42**, 7989 (1990)
385. D. Reznik, L. Pintschovius, M. Ito, S. Iikubo, M. Sato, H. Goka, M. Fujita, K. Yamada, G.D. Gu, J.M. Tranquada, Nature **440**, 1170 (2006)
386. H. Rietschel, J. Low Temp. Phys. **95**, 293 (1994)
387. O. Rösch, O. Gunnarson, X. Zhou, T. Yoshida, T. Sasagawa, A. Fujimori, Z. Hussain, Z.X. Shen, S. Uchida, Phys. Rev. Lett. **95**, 227002 (2005)

388. T.R. Sendyka, W. Dmowski, T. Egami, N. Seiji, H. Yamauchi, S. Tanaka, Phys. Rev. B **51**, 6747 (1995)
389. R. Zamboni, A.J. Pal, C. Taliani, Solid State Commun. **70**, 813 (1989)
390. G.-M. Zhao, K. Conder, H. Keller, K.A. Müller, Nature (London) **381**, 676 (1996)
391. T. Egami, in *Polarons in Bulk Materials and Systems with Reduced Dimensionality, International School of Physics Enrico Fermi*, vol. 7, ed. by G. Iadonisi, J. Ranninger, G.D. Filipis (IOS Press, Amsterdam, 2006), p. 101
392. J.C. Phillips, A.R. Bishop, A. Saxena, Rep. Prog. Phys. **66**, 2111 (2003)
393. Y. Shinozuka, Y. Toyozawa, J. Phys. Soc. Jpn. **46**, 505 (1979)
394. Y. Toyozawa, Physica B **116**, 7 (1983)
395. F.X. Bronold, A. Saxena, A.R. Bishop, Phys. Rev. B **63**, 235109 (2001)
396. A.R. Bishop, J. Phys. Chem. Solids **65**, 1449 (2004)
397. J.C. Phillips, J. Jung, Philos. Mag. B **82**, 1163 (2002)
398. F.C. Zhang, T.M. Rice, Phys. Rev. B **37**, 3759 (1989)
399. J. Zhong, H.B. Schuttler, Phys. Rev. Lett. **69**, 1600 (1992)
400. J.K. Freericks, G.D. Mahan, Phys. Rev. B **54**, 9372 (1996)
401. H. Fehske, D. Ihle, J. Loos, U. Trapper, H. Buttner, Z. Phys. B **94**, 91 (1994)
402. T. Sakai, D. Poilblanc, D.J. Scalapino , Phys. Rev. B **55**, 8445 (1997)
403. K. Yonemitsu, J. Zhong, H.B. Schuttler, Phys. Rev. B **59**, 1444 (1999)
404. A.S. Mishchenko, N. Nagaosa, Phys. Rev. Lett. **93**, 036402 (2004)
405. P. Prelovsek, R. Zeyher, P. Horsch, Phys. Rev. Lett. **96**, 086402 (2006)
406. J. Riera, A. Moreo, Phys. Rev. B **73**, 014518 (2006)
407. Z.B. Huang, W. Hanke, E. Arrigoni, J. Scalapino, Phys. Rev. B **68**, 220507 (2003)
408. W. Koller, D. Meyer, A.C. Hewson, Phys. Rev. B **70**, 155103 (2004)
409. B.J. Alder, K.J. Runge, R.T. Scalettar, Phys. Rev. Lett. **79**, 3022 (1997)
410. Z. Radovic, N. Bozovic, I. Bozovic, Phys. Rev. B **77**, 092508 (2008)
411. Y. Tokura,*Colossal Magnetoresistance Oxides*(Gordon and Breach, New York, 2000)
412. V.L. Ginzburg, Contemp. Phys. **9**, 355 (1968)
413. W.A. Little, Phys. Rev. A **134**, 1416 (1964)
414. H. Fröhlich, Phys. Lett. A **26**, 169 (1968)
415. E.A. Pashitskii, Zh. Eksp. Teor. Fiz. **55**, 2387 (1968) [Sov. Phys. JETP **28**, 1267 (1969)]
416. A.J. Millis, H. Monien, D. Pines, Phys. Rev. B **42**, 167 (1990)
417. J.R. Schrieffer, X.G. Wen, S.C. Zhang, Phys. Rev. B **39**, 11663 (1989)
418. J.E. Hirsch, Phys. Rev. B **62**, 14487 (2002)
419. V.I. Belyavsky, Yu.V. Kopaev, Phys. Rev. B **67**, 024513 (2003)
420. P.W. Anderson, P.A. Lee, M. Randeria, T.M. Rice, N. Trivedi, F.C. Zhang, J. Phys. Condens. Matter **16**, R755 (2004)
421. W. Kohn, J.M. Luttinger, Phys. Rev. Lett. **15**, 524 (1965)
422. F. Becca, M. Tarquini, M. Grilli, C. Di Castro, Phys. Rev. B **54**, 12443 (1996)
423. A.M. Gabovich, A.I. Voitenko, J.F. Annett, M. Ausloos, Supercond. Sci. Technol. **14**, R1 (2001)
424. P.W. Anderson, Science **316**, 1705 (2007)
425. T. Aimi, M. Imada, J. Phys. Soc. Jpn. **76**, 113708 (2007)
426. N. Furukawa, M. Imada, Ann. Phys. **61**, 3331 (1992)
427. S. Zhang, J. Carlson, J.E. Gubernatis, Phys. Rev. Lett. **78**, 4486 (1997)

428. J.G. Bednorz, K.A. Müller, Z. Phys. B **64**, 189 (1986)
429. P.P. Edwards, C.N.R. Rao, N. Kumar, A.S. Alexandrov, ChemPhysChem **7**, 2015 (2006)
430. J.H. Kim, B.J. Feenstra, H.S. Somal, D. van der Marel, W.Y. Lee, A.M. Gerrits, A. Wittlin, Phys. Rev. B **49**, 13065 (1994)
431. P.B. Allen, Nature **412**, 494 (2001)
432. J. Lee, K. Fujita, K. McElroy, J.A. Slezak, M. Wang, Y. Aiura, H. Bando, M. Ishikado, T. Masui, J.X. Zhu, A.V. Balatsky, H. Eisaki, S. Uchida, J.C. Devis, Nature **442**, 546 (2006)
433. A.S. Alexandrov, Physica C (Amsterdam) **363**, 231 (2001)
434. D. Mihailović, V.V. Kabanov, K. Zagar, J. Demsar, Phys. Rev. B **60**, 6995 (1999)
435. T. Timusk, B. Statt, Rep. Prog. Phys. **62**, 61 (1999)
436. V.N. Zavaritsky, V.V. Kabanov, A.S. Alexandrov, Europhys. Lett. **60**, 127 (1998)
437. A.S. Alexandrov, Phys. Rev. B **48**, 10571 (1993)
438. A.S. Alexandrov, N.F. Mott, Phys. Rev. Lett. **71**, 1075 (1993)
439. Y. Zhang, N.P. Ong, Z.A. Xu, K. Krishana, R. Gagnon, L. Taillefer, Phys. Rev. Lett. **84**, 2219 (2000)
440. K.K. Lee, A.S. Alexandrov, W.Y. Liang, Phys. Rev. Lett. **90**, 217001 (2003)
441. C. Bergemann, A.W. Tyler, A.P. Mackenzie, J.R. Cooper, S.R. Julian, D.E. Farrell, Phys. Rev. B **57**, 14387 (1998)
442. Y. Wang, L. Li, M.J. Naughton, G.D. Gu, S. Uchida, P. Ong Phys. Rev. Lett. **95**, 247002 (2005)
443. A.S. Alexandrov, Phys. Rev. Lett. **96**, 147003 (2006)
444. M.B. Salamon, M. Jaime, Rev. Mod. Phys. **73**, 583 (2001)
445. P.W. Anderson, H. Hasegawa, Phys. Rev. **100**, 675 (1955)
446. C. Zener, Phys. Rev. **82**, 403 (1951)
447. A.J. Millis, Nature **392**, 147 (1998)
448. A.J. Millis, B.I. Shraiman, R. Mueller, Phys. Rev. Lett. **77**, 175 (1996)
449. B. Michaelis, A.J. Millis, Phys. Rev. B **68**, 115111 (2003)
450. G.M. Zhao, in *Polarons in Advanced Materials*, ed. by A.S. Alexandrov (Canopus/Springer, Bristol, 2007), p. 569
451. M. Jaime, H.T. Hardner, M.B. Salamon, M. Rubinstein, P. Dorsey, D. Emin, Phys. Rev. Lett. **78**, 951 (1997)
452. S. Yunoki, J. Hu, A.L. Malvezzi, A. Moreo, N. Furukawa, E. Dagotto, Phys. Rev. Lett. **80**, 845 (1998)
453. A. Moreo, M. Mayr, S. Yunoki, E. Dagotto, Phys. Rev. Lett. **84**, 5568 (2000)
454. J. Burgy, A. Moreo, E. Dagotto, Phys. Rev. Lett. **92**, 097202 (2004)
455. K.H. Kim, J.H. Jung, T.W. Noh, Phys. Rev. Lett. **81**, 1517 (1977)
456. Y. Okimoto, T. Katsufuji, T. Ishikawa, T. Arima, Y. Tokura, Phys. Rev. B **55**, 4206 (1997)
457. H.L. Ju, H.C. Sohn, K.M. Krishnan, Phys. Rev. Lett. **79**, 3230 (1997)
458. D.S. Dessau, T. Saitoh, C.H. Park, Z.X. Shen, Y. Moritomo, Y. Tokura, Int. J. Mod. Phys. B **12**, 3389 (1998)
459. J.-S. Zhou, J.B. Goodenough, A. Asamitsu, Y. Tokura, Phys. Rev. Lett. **79**, 3234 (1997)
460. T. Saitoh, A.E. Bocquet, T. Mizokawa, H. Namatame, A. Fujimori, M. Abbate, Y. Takeda, M. Takano, Phys. Rev. B **51**, 13942 (1995)

461. M.A. Subramanian, B.H. Toby, A.P. Ramirez, W.J. Marshall, A.W. Sleight, G.H. Kwei, Science **273**, 81 (1996)
462. G.M. Zhao, Phys. Rev. B **62**, 11 639 (2001)
463. S.J.L. Billinge, R.G. DiFrancesco, G.H. Kwei, J.J. Neumeier, J.D. Thompson, Phys. Rev. Lett. **77**, 715 (1996)
464. P. Dai, J.A. Fernandez-Baca, N. Wakabayashi, E.W. Plummer, Y. Tomioka, Y. Tokura, Phys. Rev. Lett. **85**, 2553 (2000)
465. S. Shimomura, N. Wakabayashi, H. Kuwahara, Y. Tokura, Phys. Rev. Lett. **83**, 4389 (1999)
466. L. Vasiliu-Doloc, G. Rosenkranz, R. Osborn, S.K. Sinha, J.W. Lynn, J. Mesot, O.H. Seeck, G. Preosti, A.J. Fedro, J.F. Mitchell, Phys. Rev. Lett. **83**, 4393 (1999)
467. D. Louca, T. Egami, E.L. Brosha, H. Röder, A.R. Bishop, Phys. Rev. B **56**, R8475 (1997)
468. A. Nucara, A. Perucchi, P. Calvani, T. Aselage, D. Emin, Phys. Rev. B **68**, 174432 (2003)
469. A.S. Alexandrov, A.M. Bratkovsky, V.V. Kabanov, Phys. Rev. Lett. **96**, 117003 (2006)
470. M. Egilmez, K.H. Chow, J. Jung, Z. Salman, Appl. Phys. Lett. **90**, 162508 (2007)
471. L.M. Wang, H.C. Yang, H.E. Horng, Phys. Rev. B **64**, 224423 (2001)
472. W. Westerburg, F. Martin, P.J.M. van Bentum, J.A.A.J. Perenboom, G. Jakob, Europhys. J.B **14**, 509 (2000)
473. J.E. Gordon, C. Marcenat, J.P. Franck, I. Isaac, G.W. Zhang, R. Lortz, C. Meingast, F. Bouquet, R.A. Fisher, N.E. Phillips, Phys. Rev. B **65**, 024441 (2002)
474. P. Velasco, J. Mira, F. Guinea, J. Rivas, M.J. Martinez-Lope, J.A. Alonso, J.L. Martinez, Phys. Rev. B **66**, 104412 (1996)
475. N. Mannella, W.L. Yang, K. Tanaka, X.J. Zhou, H. Zheng, J.F. Mitchell, J. Zaanen, T.P. Devereaux, N. Nagaosa, T.P. Devereaux, N. Nagaosa, Z. Hussain Z.X. Shen, Phys. Rev. B **76**, 233102 (2007)
476. N. Mannella, W. Yang, X.J. Zhou, H. Zheng, J.F. Mitchell, J. Zaanen, T.P. Devereaux, N. Nagaosa, T.P. Devereaux, N. Nagaosa, Z. Hussain Z.X. Shen, Nature **438**, 474 (2005)
477. A.S. Alexandrov, Physica C (Amsterdam) **341–348**, 107 (2000)
478. E.L. Nagaev, Phys. Rev. Lett. **84**, 2042 (2000)
479. E.L. Nagaev, Phys. Rep. **346**, 388 (2001)
480. A.S. Alexandrov, in *Nanotechnology for Electronic Materials and Devices Series: Nanostructure Science and Technology*, ed. by A. Korkin, E. Gusev, J.K. Labanowski, S. Luryi (Springer, New York, 2006), p. 305
481. J.T. Devreese, V.M. Fomin E.P. Pokatilov, in *Handbook of Semiconductor Nanostructures and Nanodevices*, vol. 4, ed. by A.A. Balandin, K.L. Wang (American Scientific Publishers, Los Angeles, 2006), p. 339
482. M. Galperin, M.A. Ratner, A. Nitzan, J. Phys. Condens. Matter **19**, 103201 (2007)
483. V.M. Fomin, V.N. Gladilin, J.T. Devreese, E.P. Pokatilov, S.N. Balaban, S.N. Klimin, Phys. Rev. B **57**, 2415 (1998)
484. V.A. Fonoberov, E.P. Pokatilov, V.M. Fomin, J.T. Devreese, Phys. Rev. Lett. **92**, 127402 (2004)

485. E.P. Pokatilov, S.N. Klimin, V.M. Fomin, J.T. Devreese, F.W. Wise, Phys. Rev. B **65**, 075316 (2002)
486. A. García-Cristóbal, A.W.E. Minnaert, V.M. Fomin, J.T. Devreese, A.Y. Silov, J.E.M. Haverkort, J.H. Wolter, Physica Status Solidi (b) **215**, 331 (1999)
487. A. Lemaître, A.D. Ashmore, J.J. Finley, D.J. Mowbray, M.S. Skolnick, M. Hopkinson, T.F. Krauss, Phys. Rev. B **63**, 161309(R) (2001)
488. A. Zrenner, F. Findeis, M. Baier, M. Bichler, G. Abstreiter, Physica B **298**, 239 (2001)
489. R. Heitz, H. Born, F. Guffarth, O. Stier, A. Schliwa, A. Hoffmann, D. Bimberg, Phys. Rev. B **64**, 241305(R) (2001)
490. U. Woggon, D. Miller, F. Kalina, B. Gerlach, D. Kayser, K. Leonardi, D. Kayser D. Hommel, Phys. Rev. B **67**, 045204 (2003) (pages 5)
491. S. Mukhopadhyay, A. Chatterjee, J. Phys. Condens. Matter **8**, 4017 1996
492. E.P. Pokatilov, V.M. Fomin, J.T. Devreese, S.N. Balaban, S.N. Klimin, J. Phys. Condens. Matter **11**, 9033 (1999)
493. R.T. Senger, A. Ercelebi, Europhys. J. B **16**, 439 (2000)
494. M. Hohenadler, P.B. Littlewood, Phys. Rev. B **76**, 155122 (2007)
495. F. Brosens, J.T. Devreese, L.F. Lemmens, Phys. Rev. E **55**, 6795 (1997)
496. F. Brosens, J.T. Devreese, L.F. Lemmens, Phys. Rev. E **55**, 227 (1997)
497. F. Brosens, J.T. Devreese, L.F. Lemmens, Phys. Rev. E **58**, 1634 (1998)
498. L.F. Lemmens, F. Brosens, J.T. Devreese, Phys. Rev. E **53**, 4467 (1996)
499. S.N. Klimin, V.M. Fomin, F. Brosens, J.T. Devreese, Phys. Rev. B **69**, 235324 (2004)
500. S.N. Klimin, F. Brosens, J.T. Devreese, (2006), Phys. Rev. B **77**, 085308 (2008)
501. M. Hohenadler, P.B. Littlewood, H. Fehske, Phys. Rev. B **76**, 184303 (2007)
502. N.-B. Zhitenev, H. Meng, Z. Bao, Phys. Rev. Lett. **88**, 226801 (2002)
503. H. Park, J. Park, A.K.L. Lim, E.H. Anderson, A.P. Alivisatos, P.L. McEuen, Nature **407**, 57 (2000)
504. A.S. Alexandrov, A.M. Bratkovsky, Phys. Rev. B **67**, 235312 (2003)
505. V.N. Ermakov, Physica E **8**, 99 (2000)
506. L.I. Glazman, R.I. Shekhter, Zh. Eksp. Teor. Fiz. **94**, 292 (1988) [Sov. Phys. JETP **67**, 163]
507. U. Lundin, R.H. McKenzie, Phys. Rev. B **66**, 075303 (2002)
508. H. Ness, S.A. Shevlin, A.J. Fisher, Phys. Rev. B **63**, 125422 (2001)
509. N.S. Wingreen, K.W. Jacobsen, J.W. Wilkins, Phys. Rev. B **40**, 11834 (1989)
510. A.S. Alexandrov, A.M. Bratkovsky, R.S. Williams, Phys. Rev. B **67**, 075301 (2003)
511. A. Mitra, I. Aleiner, A.J. Millis, Phys. Rev. Lett. **94**, 076404 (2005)
512. D. Mozyrsky, M.B. Hastings, I. Martin, Phys. Rev. B **73**, 035104 (2006)
513. J. Koch, M.E. Raikh, F. von Oppen, Phys. Rev. Lett. **96**, 056803 (2006)
514. Z.K. Keane, J.W. Ciszek, J.M. Tour, D. Natelson, Nano Lett. **6**, 1518 (2006)
515. A.M. Bratkovsky, in *Polarons in Advanced Materials*, ed. by A.S. Alexandrov (Canopus/Springer, Bristol, 2007), p. 623
516. J.T. Devreese, in *Encyclopedia of Modern Optics*, 5 vols., ed. by B.D. Guenther, D.G. Steel, L. Bayvel (Academic, London, 2004), p. 629
517. A. Macridin, B. Moritz, M. Jarrell, T. Maier, Phys. Rev. Lett. **97**, 056402 (2006)
518. N.F. Mott, in *Polarons and Bipolarons in High Tc Superconductors and Related Materials*, ed. by E.K.H. Salje, A.S. Alexandrov, W.Y. Liang (Cambridge University Press, Cambridge, 1995), p. 1
519. A.L. Shluger, A.M. Stoneham, J. Phys. Condens. Matter **1**, 3049 (1993)

Index